Convex Optimization

Series Editor
Nikolaos Limnios

Convex Optimization

Introductory Course

Mikhail Moklyachuk

iSTE

WILEY

First published 2020 in Great Britain and the United States by ISTE Ltd and John Wiley & Sons, Inc.

ISTE Ltd
27-37 St George's Road
London SW19 4EU
UK

www.iste.co.uk

John Wiley & Sons, Inc.
111 River Street
Hoboken, NJ 07030
USA

www.wiley.com

Library of Congress Control Number: 2020943973

British Library Cataloguing-in-Publication Data
A CIP record for this book is available from the British Library
ISBN 978-1-78630-683-8

Contents

Notations

$\mathbb{N}$	Set of natural numbers
$\mathbb{Z}$	Set of integer numbers
$\mathbb{Z}_+$	Set of non-negative integer numbers
$\mathbb{R}$	Set of real numbers
$\overline{\mathbb{R}} = \mathbb{R} \cup \{+\infty\}$	Extended set of real numbers
$\mathbb{Q}$	Set of rational numbers
$\mathbb{R}^n$	Set of real n-vectors
$\mathbb{R}^{m\times n}$	Set of real $m \times n$-matrices
$\mathbb{R}_+$	Set of non-negative real numbers
$\mathbb{R}_{++}$	Set of positive real numbers
$\mathbb{C}$	Set of complex numbers
$\mathbb{C}^n$	Set of complex n-vectors
$\mathbb{C}^{m\times n}$	Set of complex $m \times n$-matrices
$\mathbb{S}^n$	Set of symmetric $n \times n$-matrices
$\mathbb{S}^n_+$	Set of symmetric positive semidefinite $n \times n$-matrices
$\mathbb{S}^n_{++}$	Set of symmetric positive definite $n \times n$-matrices
$\mathbb{I}$	Identity matrix
$X^\top$	Transpose of matrix X
$\mathrm{tr}\,(X)$	Trace of matrix X
$\lambda_i(X)$	ith largest eigenvalue of symmetric matrix X
$\langle \cdot\,,\,\cdot \rangle$	Inner product
$x \perp y$	Vectors x and y are orthogonal: $\langle x, y \rangle = 0$
$V^\perp$	Orthogonal complement of subspace V
$\mathrm{diag}(X)$	Diagonal matrix with diagonal entries $x_1, \ldots, x_n$
$\mathrm{rank}\,(X)$	Rank of matrix X
$\|\cdot\|$	A norm
$\|\cdot\|_*$	Dual of norm $\|\cdot\|$
$\|x\|_2$	Euclidean norm of vector x
$x \preceq y$	Componentwise inequality between vectors x and y

$x \prec y$	Strict componentwise inequality between vectors x and y
$X \preceq Y$	Matrix inequality between symmetric matrices X and Y
$X \prec Y$	Strict matrix inequality between symmetric matrices X and Y
$X \preceq_K Y$	Generalized inequality induced by proper cone K
$X \prec_K Y$	Strict generalized inequality induced by proper cone K
$\operatorname{int} X$	Interior of set X
$\operatorname{ri} X$	Relative interior of set X
$\operatorname{conv} X$	Convex hull of set X
$\operatorname{aff} X$	Affine hull of set X
$\operatorname{cone} X$	Conic hull of set X
$\operatorname{Lin} X$	Linear hull of set X
$\overline{X}$	Closure of set X
$\overline{\operatorname{conv}}\, X$	Closed convex hull of set X
$\dim X$	Dimension of set X
∂X	Boundary of set X
K^*	Dual cone associated with cone K
$l^+_{\hat{x}h}$	A ray proceeding from a point $\hat{x}$ in the direction h
$H_{p\beta}$	A hyperplane with the normal vector p
$H^+_{p\beta}$, $H^-_{p\beta}$	Half-spaces generated by hyperplane $H_{p\beta}$
$\pi_X(a)$	Projection of point a onto set X
$\rho(X_1, X_2)$	Distance between sets X_1 and X_2
$\operatorname{epi} f$	Epigraph of function f
$S_r(f)$	Sublevel set of function f
$\operatorname{dom} f$	Effective set of function f
$f_1 \oplus f_2$	Infimal convolution of functions f_1, f_2
$\mu(x\|X)$	Minkowski function
$\gamma_X(x)$	Gauge function
$\delta(x\|X)$	Indicator function
$\sigma(x\|X)$	Support function
f^*	Conjugate function
$\partial f(x)$	Subdifferential of function f at point x
$\overline{\partial} f(x)$	Superdifferential of function f at point x
$\Pi(\mathbb{R}^m)$	Set of all non-empty subsets of the space $\mathbb{R}^m$

Introduction

Convex analysis and optimization have an increasing impact on many areas of mathematics and applications including control systems, estimation and signal processing, communications and networks, electronic circuit design, data analysis and modeling, statistics, and economics and finance. There are several fundamental books devoted to different aspects of convex analysis and optimization. Among them we can mention *Optima and Equilibria: An Introduction to Nonlinear Analysis* by Aubin (1998), *Convex Analysis* by Rockafellar (1970), *Convex Analysis and Minimization Algorithms* (in two volumes) by Hiriart-Urruty and Lemaréchal (1993) and its abridged version (2002), *Convex Analysis and Nonlinear Optimization* by Borwein and Lewis (2000), *Convex Optimization* by Boyd and Vandenberghe (2004), *Convex Analysis and Optimization* by Bertsekas *et al.* (2003), *Convex Analysis and Extremal Problems* by Pshenichnyj (1980), *A Course in Optimization Methods* by Sukharev *et al.* (2005), *Convex Analysis: An Introductory Text* by Van Tiel (1984), as well as other books listed in the bibliography (see Alekseev *et al.* (1984, 1987); Alekseev and Timokhov (1991); Clarke (1983); Hiriart-Urruty (1998); Ioffe and Tikhomirov (1979) and Nesterov (2004)).

This book provides easy access to the basic principles and methods for solving constrained and unconstrained convex optimization problems. Structurally, the book has been divided into the following parts: basic methods for solving constrained and unconstrained optimization problems with differentiable objective functions, convex sets and their properties, convex functions, their properties and generalizations, subgradients and subdifferentials, and basic principles and methods for solving constrained and unconstrained convex optimization problems. The first part of the book describes methods for finding the extremum of functions of one and many variables. Problems of constrained and unconstrained optimization (problems with restrictions of inequality and inequality types) are investigated. The necessary and sufficient conditions of the extremum, the Lagrange method, are described.

The second part is the most voluminous in terms of the amount of material presented. Properties of convex sets and convex functions directly related to extreme problems are described. The basic principles of subdifferential calculus are outlined. The third part is devoted to the problems of mathematical programming. The problem of convex programming is considered in detail. The Kuhn–Tucker theorem is proved and the economic interpretations of the Kuhn–Tucker vector are described.

We give detailed proofs for most of the results presented in the book and also include many figures and exercises for better understanding of the material. Finally, we present solutions and hints to selected exercises at the end of the book. Exercises are given at the end of each chapter while figures and examples are provided throughout the whole text. The list of references contains texts which are closely related to the topics considered in the book and may be helpful to the reader for advanced studies of convex analysis, its applications and further extensions. Since only elementary knowledge in linear algebra and basic calculus is required, this book can be used as a textbook for both undergraduate and graduate level courses in convex optimization and its applications. In fact, the author has used these lecture notes for teaching such classes at Kyiv National University. We hope that the book will make convex optimization methods more accessible to large groups of undergraduate and graduate students, researchers in different disciplines and practitioners. The idea was to prepare materials of lectures in accordance with the suggestion made by Einstein: "Everything should be made as simple as possible, but not simpler."

1

Optimization Problems with Differentiable Objective Functions

1.1. Basic concepts

The word "maximum" means the largest, and the word "minimum" means the smallest. These two concepts are combined with the term "extremum", which means the extreme. Also pertinent is the term "optimal" (from Latin *optimus*), which means the best. The problems of determining the largest and smallest quantities are called *extremum problems*. Such problems arise in different areas of activity and therefore different terms are used for the descriptions of the problems. To use the theory of extremum problems, it is necessary to describe problems in the language of mathematics. This process is called the formalization of the problem.

The *formalized problem* consists of the following elements:

– objective function $f\colon X \to \bar{\mathbb{R}}$;

– domain X of the definition of the objective functional f;

– constraint: $C \subset X$.

Here, $\bar{\mathbb{R}}$ is an extended real line, that is, the set of all real numbers, supplemented by the values $+\infty$ and $-\infty$, C is a subset of the domain of definition of the objective functional f. So to formalize an optimization problem means to clearly define and describe elements f, C and X. The formalized problem is written in the form

$$f(x) \to \inf\ (\sup), \quad x \in C. \tag{1.1}$$

Points of the set C are called *admissible points* of the problem [1.1]. If $C = X$, then all points of the domain of definition of the function are admissible. The problem [1.1] in this case is called a *problem without constraints*.

The maximization problem can always be reduced to the minimization problem by replacing the functional f with the functional $g = -f$. And, on the contrary, the minimization problem in the same way can be reduced to the maximization problem. If the necessary conditions for the extremum in the minimization problem and maximization problem are different, then we write these conditions only for the minimization problem. If it is necessary to investigate both problems, then we write down

$$f(x) \to \text{extr}, \quad x \in C.$$

An admissible point $\hat{x}$ is a point of *absolute* or *global minimum* (*maximum*) of the extremum problem if for any $x \in C$ the inequality holds true

$$f(x) \geq f(\hat{x}) \quad (f(x) \leq f(\hat{x})).$$

Then we write $\hat{x} \in \text{absmin (absmax)}$. The point of the absolute minimum (maximum) is called a *solution of the problem*. The value $f(\hat{x})$, where $\hat{x}$ is a solution of the problem, is called a *numerical value* of the problem. This value is denoted by $S_{\min}$ ($S_{\max}$).

In addition to global extremum problems, local extremum problems are also studied. Let X be a normed space. A *local minimum* (*maximum*) of the problem is reached at a point $\hat{x}$, that is $\hat{x} \in \text{locmin (locmax)}$, if $\hat{x} \in C$ and there exists a number $\delta > 0$ such that for any admissible point $x \in C$ that satisfies condition $\|x - \hat{x}\| < \delta$, the inequality holds true

$$f(x) \geq f(\hat{x}) \quad (f(x) \leq f(\hat{x})).$$

In other words, if $\hat{x} \in \text{locmin (locmax)}$, then there exists a neighborhood $O_{\hat{x}}$ of the point $\hat{x}$ such that

$$\hat{x} \in \text{absmin (absmax)}$$

in the problem

$$f(x) \to \inf\ (\sup), \quad x \in C \cap O_{\hat{x}}.$$

The theory of extremum problems gives general rules for solving extremum problems. The theory of necessary conditions of the extremum is more developed. The necessary conditions of the extremum make it possible to allocate a set of points among which solutions of the problem are situated. Such a set is called a *critical* set, and the points themselves are called *critical* points. As a rule, a critical set does not contain many points and a solution of the problem can be found by one or another method.

1.2. Optimization problems with objective functions of one variable

Let $f\colon \mathbb{R} \to \mathbb{R}$ be a function of one real variable.

DEFINITION 1.1.– *A function f is said to be lower semicontinuous (upper semicontinuous) at a point $\hat{x}$ if for every $\varepsilon > 0$, there exists a $\delta > 0$ such that the inequality*

$$f(x) > f(\hat{x}) - \varepsilon \quad (f(x) < f(\hat{x}) + \varepsilon)$$

holds true for all $x \in (\hat{x} - \delta, \hat{x} + \delta)$.

DEFINITION 1.2.– *(Equivalent) A function f is said to be lower semicontinuous (upper semicontinuous) at a point $\hat{x}$ if for every $a \in \mathbb{R}$, $a < f(\hat{x})$ $(a > f(\hat{x}))$ there exists $\delta > 0$, such that the inequality*

$$f(x) > a \quad (f(x) < a)$$

holds true for all $x \in (\hat{x} - \delta, \hat{x} + \delta)$.

If the function takes values in $\bar{\mathbb{R}} = \mathbb{R} \cup \{-\infty\} \cup \{+\infty\}$, then definitions 1.1 and 1.2 make sense when $f(\hat{x}) = +\infty$ $(f(\hat{x}) = -\infty)$. In the case where $f(\hat{x}) = -\infty$ $(f(\hat{x}) = +\infty)$, the function f is considered to be *lower semicontinuous* (*upper semicontinuous*) by agreement.

Here are examples of semicontinuous functions:

1) the function $y = [x]$ (integer part of x) is upper semicontinuous at the points of discontinuity;

2) the function $y = \{x\}$ (fractional part of x) is lower semicontinuous at the points of discontinuity;

3) the Dirichlet function, which is equal to 0 at rational points and equal to 1 at irrational points, is lower semicontinuous at each rational point and upper semicontinuous at each irrational point;

4) if the function $f\colon \mathbb{R} \to \bar{\mathbb{R}}$ has a local minimum (maximum) at the point $\hat{x}$, then it is lower (upper) semicontinuous at the point ;

5) the function $f(x) = \frac{1}{|x|}$ for $x \neq 0$, $f(0) = +\infty$, is lower semicontinuous at the point 0. If we define the function at the point 0 arbitrarily, then it will remain lower semicontinuous.

THEOREM 1.1.– Let f and g be lower semicontinuous functions. Then:

– the function $f + g$ is lower semicontinuous;

– the function αf is lower semicontinuous for $\alpha \geq 0$ and it is upper semicontinuous for $\alpha \leq 0$;

– the function $f \cdot g$ is lower semicontinuous for $f \geq 0$, $g \geq 0$;

– the function $1/f$ is upper semicontinuous if $f > 0$;

– the function $\max\{f, g\}$, $\min\{f, g\}$ is lower semicontinuous;

– the functions $\sup\{f_i\}$ $(\inf\{f_i\})$ are lower (upper) semicontinuous, if the functions f_i are lower (upper) semicontinuous.

THEOREM 1.2.– (*Weierstrass theorem*) A lower (upper) semicontinuous on the interval $[a, b]$ function $f\colon \mathbb{R} \to \mathbb{R}$ is bounded from below (from above) on $[a, b]$ and attains the smallest (largest) value.

THEOREM 1.3.– (*Fermat's theorem*) If $\hat{x}$ is a point of local extremum of the differentiable at the point $\hat{x}$ function $f(x)$, then $f'(\hat{x}) = 0$.

Fermat's theorem gives the first-order necessary condition for existence of a local extremum of the function $f(x)$ at point $\hat{x}$. The following theorems give the second-order necessary and sufficient conditions for the extremum.

THEOREM 1.4.– (*Necessary conditions of the second order*) If $\hat{x}$ is a point of local minimum (maximum) of the function $f(x)$, which has the second-order derivative at the point $\hat{x}$, then the following conditions hold true:

$$f'(\hat{x}) = 0, \quad f''(\hat{x}) \geq 0 \quad (f''(\hat{x}) \leq 0).$$

THEOREM 1.5.– (*Sufficient conditions of the second order*) If the function $f(x)$ has at a point $\hat{x}$ the second-order derivative and

$$f'(\hat{x}) = 0, \quad f''(\hat{x}) > 0 \quad (f''(\hat{x}) < 0),$$

then $\hat{x}$ is the point of local minimum (maximum) of the function $f(x)$.

The necessary and sufficient conditions of the higher order of existence of an extremum of the function $f(x)$ are given in the following theorems.

THEOREM 1.6.– (*Necessary conditions of higher order*) If $\hat{x}$ is a point of local minimum (maximum) of the function $f(x)$, which has at this point $\hat{x}$ the nth order derivative, then

$$f'(\hat{x}) = \ldots = f^{(n)}(\hat{x}) = 0,$$

or

$$f'(\hat{x}) = \ldots = f^{(2m-1)}(\hat{x}) = 0, \; f^{(2m)}(\hat{x}) > 0 \quad (f^{(2m)}(\hat{x}) < 0)$$

for some $m \geq 1$, $2m \leq n$.

PROOF.– According to Taylor's theorem, for a function which is n times differentiable at the point $\hat{x}$, we have

$$f(\hat{x}+x)=\sum_{k=0}^{n}\frac{f^{(k)}(\hat{x})}{k!}x^k+r(x),\quad \frac{r(x)}{x^n}\to 0,\quad \text{as}\quad x\to 0.$$

If $n=1$, then the assertion of the theorem is true as a result of Fermat's theorem. Let $n>1$, then

$$f'(\hat{x})=\ldots=f^{(n)}(\hat{x})=0$$

or

$$f'(\hat{x})=\ldots=f^{(l-1)}(\hat{x})=0,\quad f^{(l)}(\hat{x})\neq 0,\quad l\leq n.$$

Let l be an odd number. Then the function $g(u)=f(\hat{x}+u^{\frac{1}{l}})$, $u\in R$ can be expanded in a series by Taylor's theorem

$$g(u)=f(\hat{x})+\sum_{k=l}^{n}\frac{f^{(k)}(\hat{x})}{k!}u^{\frac{k}{l}}+r(u^{\frac{1}{l}}),$$

$$r(u^{\frac{1}{l}})/u^{\frac{n}{l}}\to 0\quad \text{as}\quad u\to 0.$$

The function $g(u)$ has a derivative at point $u=0$. Since $\hat{x}\in\operatorname{locmin} f$, then $0\in\operatorname{locmin} g$. According to Fermat's theorem, $g'(0)=f^{(l)}(\hat{x})/l!=0$. Hence $f^{(l)}(\hat{x})=0$. This contradicts the condition $f^{(l)}(\hat{x})\neq 0$. Therefore, the number l is even, $l=2m$. According to Taylor's theorem

$$f(\hat{x}+x)-f(\hat{x})=\frac{f^{(2m)}(\hat{x})x^{2m}}{(2m)!}+r_1(x),\quad \frac{r_1(x)}{x^{2m}}\to 0,\quad \text{as}\quad x\to 0.$$

Since $f^{(2m)}(\hat{x})\neq 0$, then $f^{(2m)}(\hat{x})>0$ if $\hat{x}\in\operatorname{locmin} f$ and $f^{(2m)}(\hat{x})<0$ if $\hat{x}\in\operatorname{locmax} f$. □

THEOREM 1.7.– (*Sufficient conditions of higher order*) If the function $f(x)$ has a derivative of order n at point $\hat{x}$ and

$$f'(\hat{x})=\ldots=f^{(2m-1)}(\hat{x})=0,\ f^{(2m)}(\hat{x})>0\quad (f^{(2m)}(\hat{x})<0)$$

for some $m\geq 1$, $2m\leq n$, then the function $f(x)$ attains a local minimum (maximum) at the point $\hat{x}$.

PROOF.– Since $f'(\hat{x})=\ldots=f^{(2m-1)}(\hat{x})=0$, then according to Taylor's theorem

$$f(\hat{x}+x)-f(\hat{x})=\frac{f^{(2m)}(\hat{x})x^{2m}}{(2m)!}+r_1(x),\quad \frac{r_1(x)}{x^{2m}}\to 0,\quad \text{as}\quad x\to 0.$$

If $f^{(2m)}(\hat{x})>0$, then $f(\hat{x}+x)-f(\hat{x})\geq 0$ for sufficiently small x, i.e. $\hat{x}\in\operatorname{locmin} f$. If $f^{(2m)}(\hat{x})<0$, then $f(\hat{x}+x)-f(\hat{x})\leq 0$ for sufficiently small x, i.e. $\hat{x}\in\operatorname{locmax} f$. □

1.3. Optimization problems with objective functions of several variables

Let $f\colon \mathbb{R}^n \to \overline{\mathbb{R}}$ be a function of n real variables.

DEFINITION 1.3.– *A function f is said to be lower semicontinuous (upper semicontinuous) at a point $\hat{x}$ if there exists a δ-neighborhood*

$$O_{\hat{x}} = \{x\colon \|x - \hat{x}\| < \delta\}, \quad \|x\| = \left(\sum_{k=1}^{n} x_k^2\right)^{\frac{1}{2}},$$

of the point $\hat{x}$ such that the inequality

$$f(x) > f(\hat{x}) - \varepsilon \quad (f(x) < f(\hat{x}) + \varepsilon)$$

holds true for all $x \in O_{\hat{x}}$.

THEOREM 1.8.– A function $f\colon \mathbb{R}^n \to \overline{\mathbb{R}}$ is lower semicontinuous on $\mathbb{R}^n$ if and only if for all $a \in \mathbb{R}$ the set $f^{-1}((a, +\infty])$ is open (or the complementary set $f^{-1}((-\infty, a])$ is closed).

PROOF.– Let f be a lower semicontinuous on $\mathbb{R}^n$ function, let $a \in \mathbb{R}$ and let $\hat{x} \in f^{-1}((a, +\infty])$. Then there exists a δ-neighborhood $O_{\hat{x}}$ of the point $\hat{x}$ such that for all points $x \in O_{\hat{x}}$ the inequality $f(x) > a$ holds true. This means that $O_{\hat{x}} \subset f^{-1}((a, +\infty])$. Consequently, the set $f^{-1}((a, +\infty])$ is open.

Vice versa, if the set $f^{-1}((a, +\infty])$ is open for any $a \in \mathbb{R}$ and $\hat{x} \in \mathbb{R}^n$, then $f(\hat{x}) = -\infty$ and the function f is lower semicontinuous at point $\hat{x}$ by agreement, or $f(\hat{x}) > -\infty$ and $\hat{x} \in f^{-1}((a, +\infty])$, when $a < f(\hat{x})$. Since the set $f^{-1}((a, +\infty])$ is open, then there exists a δ-neighborhood $O_{\hat{x}}$ of the point $\hat{x}$ such that $O_{\hat{x}} \subset f^{-1}((a, +\infty])$ and $f(x) > a$ for all $x \in O_{\hat{x}}$. Consequently, the function f is lower semicontinuous at point $\hat{x}$. □

THEOREM 1.9.– (*Weierstrass theorem*) A lower (upper) semicontinuous on a non-empty bounded closed subset X of the space $\mathbb{R}^n$ function is bounded from below (from above) on X and attains the smallest (largest) value.

THEOREM 1.10.– (*Weierstrass theorem*) If a function $f(x)$ is lower semicontinuous and for some number a the set $\{x\colon f(x) \leq a\}$ is non-empty and bounded, then the function $f(x)$ attains its absolute minimum.

COROLLARY 1.1.– If a function $f(x)$ is lower (upper) semicontinuous on $\mathbb{R}^n$ and

$$\lim_{\|x\| \to \infty} f(x) = +\infty \quad \left(\lim_{\|x\| \to \infty} f(x) = -\infty\right),$$

then the function $f(x)$ attains its minimum (maximum) on each closed subset of the space $\mathbb{R}^n$.

THEOREM 1.11.– (*Necessary conditions of the first order*) If $\hat{x}$ is a point of local extremum of the differentiable at the point $\hat{x}$ function $f(x)$, then all partial derivatives of the function $f(x)$ are equal to zero at the point $\hat{x}$:

$$\frac{\partial f(\hat{x})}{\partial x_1} = \cdots = \frac{\partial f(\hat{x})}{\partial x_n} = 0.$$

THEOREM 1.12.– (*Necessary conditions of the second order*) If $\hat{x}$ is a point of local minimum of the function $f(x)$, which has the second-order partial derivatives at the point $\hat{x}$, then the following condition holds true:

$$\sum_{k=1}^{n}\sum_{j=1}^{n} \frac{\partial^2 f(\hat{x})}{\partial x_k \partial x_j} h_k h_j \geq 0 \quad \forall\, h = (h_1, \ldots, h_n) \in \mathbb{R}^n.$$

This condition means that the matrix

$$f''(\hat{x}) = \left\{ \frac{\partial^2 f(\hat{x})}{\partial x_k \partial x_j} \right\}_{k=\overline{1,n}}^{j=\overline{1,n}}$$

is non-negative definite.

THEOREM 1.13.– (*Sufficient conditions of the second order*) Let a function $f\colon \mathbb{R}^n \to \mathbb{R}$ have the second-order partial derivatives at a point $\hat{x}$ and let the following conditions hold true:

$$\frac{\partial f(\hat{x})}{\partial x_1} = \cdots = \frac{\partial f(\hat{x})}{\partial x_n} = 0;$$

$$\sum_{k=1}^{n}\sum_{j=1}^{n} \frac{\partial^2 f(\hat{x})}{\partial x_k \partial x_j} h_k h_j > 0 \quad \forall\, h = (h_1, \ldots, h_n) \in \mathbb{R}^n,\ h \neq 0.$$

Then $\hat{x}$ is the point of local minimum of the function $f(x)$.

The second condition of the theorem means that the matrix

$$f''(\hat{x}) = \left\{ \frac{\partial^2 f(\hat{x})}{\partial x_k \partial x_j} \right\}_{k=\overline{1,n}}^{j=\overline{1,n}}$$

is positive definite.

THEOREM 1.14.– (*Sylvester's criterion*) A matrix A is positive definite if and only if its principal minors are positive. A matrix A is negative definite if and only if $(-1)^k \det A_k > 0$, where

$$A_k = \{a_{ij}\}_{i=\overline{1,k}}^{j=\overline{1,k}}, \quad k = 1, \ldots, n.$$

We write down the series of principal minors of the matrix A

$$\Delta_1 = a_{11}, \quad \Delta_2 = \begin{vmatrix} a_{11} & a_{12} \\ a_{21} & a_{22} \end{vmatrix}, \ldots, \Delta_n = \begin{vmatrix} a_{11} & \cdots & a_{1n} \\ \cdot & \cdots & \cdot \\ a_{n1} & \cdot & a_{nn} \end{vmatrix}.$$

Then:

– the matrix A is positive definite, if

$$\Delta_1 > 0,\ \Delta_2 > 0, \ldots, \Delta_n > 0;$$

– the matrix A is negative definite, if

$$\Delta_1 < 0,\ \Delta_2 > 0, \ldots, (-1)^n \Delta_n > 0;$$

– the matrix A is non-negative (non-positive) definite, if

$$\Delta_1 \geq 0,\ \Delta_2 \geq 0, \ldots, \Delta_n \geq 0\, (\Delta_1 \leq 0,\ \Delta_2 \geq 0, \ldots, (-1)^n \Delta_n \geq 0)$$

and there exists j, such that $\Delta_j = 0$;

– the matrix A is indefinite.

EXAMPLE 1.1.– Find solutions for the optimization problem

$$f(x_1, x_2) = x_1^4 + x_2^4 - (x_1 + x_2)^2 \to \text{extr}\,.$$

Solution. The function is continuous. Obviously, $S_{\max} = +\infty$. As a result of the Weierstrass theorem, the minimum is attained. The necessary conditions of the first order

$$\frac{\partial f(\hat{x})}{\partial x_1} = 0, \quad \frac{\partial f(\hat{x})}{\partial x_2} = 0,$$

are of the form

$$2x_1^3 = x_1 + x_2, \quad 2x_2^3 = x_1 + x_2.$$

Solving these equations, we find the critical points $(0,0)$, $(1,1)$, $(-1,-1)$. To use the second-order conditions, we compute matrices composed of the second-order partial derivatives:

$$A(\hat{x}) = \left\{ \frac{\partial^2 f(\hat{x})}{\partial x_k \partial x_j} \right\}_{k,j=1}^{2} = \begin{pmatrix} 12\hat{x}_1^2 - 2 & -2 \\ -2 & 12\hat{x}_2^2 - 2 \end{pmatrix},$$

$$A_1 = A(0,0) = \begin{pmatrix} -2 & -2 \\ -2 & -2 \end{pmatrix},$$

$$A_2 = A(1,1) = A(-1,-1) = \begin{pmatrix} 10 & -2 \\ -2 & 10 \end{pmatrix}.$$

The matrix $-A_1$ is non-negative definite. Therefore, the point $(0,0)$ satisfies the second-order necessary conditions for a maximum. However, a direct verification of the behavior of the function f in a neighborhood of the point $(0,0)$ shows that $(0,0) \notin \operatorname{locextr} f$. The matrix A_2 is positive definite. Thus, by theorem 1.13, the local minimum of the problem is attained at the points $(1,1)$, $(-1,-1)$.

Answer. $(0,0) \notin \operatorname{locextr}$; $(1,1), (-1,-1) \in \operatorname{locmin}$.

1.4. Constrained optimization problems

1.4.1. *Problems with equality constraints*

Let $f_k \colon \mathbb{R}^n \to \mathbb{R}$, $k = 0, 1, \ldots, m$, be differentiable functions of n real variables. *The constrained optimization problem (with equality constraints)* is the problem

$$f_0(x) \to \text{extr}, \quad f_1(x) = \ldots = f_m(x) = 0. \qquad [1.2]$$

The points $x \in \mathbb{R}^n$, which satisfy the equation $f_k(x) = 0$, $k = \overline{1,m}$, are called *admissible* in the problem [1.2]. An admissible point $\hat{x}$ gives a local minimum (maximum) of the problem [1.2] if there exists a number $\delta > 0$ such that for all admissible $x \in \mathbb{R}^n$ that satisfy the conditions $f_k(x) = 0$, $k = 1, 2, \ldots, m$, and the condition $\|x - \hat{x}\| < \delta$, the following inequality

$$f(x) \geq f(\hat{x}) \quad (f(x) \leq f(\hat{x}))$$

holds true.

The main method of solving constrained optimization problems is the *method of indeterminate Lagrange multipliers*. It is based on the fact that the solution of the constrained optimization problem [1.2] is attained at points that are critical in the unconstrained optimization problem

$$L(x, \lambda, \lambda_0) \to \text{extr},$$

where $L(x, \lambda, \lambda_0) = \sum_{k=0}^{m} \lambda_k f_k(x)$ is the *Lagrange function*, and $\lambda_0, \ldots, \lambda_m$ are the *Lagrange multipliers*.

THEOREM 1.15.– (*Lagrange theorem*) Let $\hat{x}$ be a point of local extremum of the problem [1.2], and let the functions $f_i(x)$, $i = 0, 1, \ldots, m$, be continuously differentiable in a neighborhood U of the point $\hat{x}$. Then there will exist the Lagrange multipliers $\lambda_0, \ldots, \lambda_m$, not all equal to zero, such that the stationary condition on x of the Lagrange function is fulfilled

$$L_x(\hat{x}, \lambda, \lambda_0) = 0 \Longleftrightarrow \frac{\partial L(\hat{x}, \lambda, \lambda_0)}{\partial x_j} = 0, \quad j = 1, \ldots, n.$$

In order for $\lambda_0 \neq 0$, it is sufficient that the vectors $f_1'(\hat{x}), \ldots, f_m'(\hat{x})$ be linearly independent.

To prove the theorem, we use the inverse function theorem in a finite-dimensional space.

THEOREM 1.16.– (*Inverse function theorem*) Let $F_1(x_1, \ldots, x_s), \ldots, F_s(x_1, \ldots, x_s)$ be s continuously differentiable in a neighborhood U of the point $\hat{x} = (\hat{x}_1, \ldots, \hat{x}_s)$ functions of s variables and let the Jacobian

$$\det\left\{\frac{\partial F_i(\hat{x})}{\partial x_j}\right\}_{i,j=1}^{s}$$

be not equal to zero. Then there exist numbers $\varepsilon > 0$, $\delta > 0$, $K > 0$ such that for any $y = (y_1, \ldots, y_s)$, $\|y\| \leq \varepsilon$ we can find $x = (x_1, \ldots, x_s)$, which satisfies conditions $\|x\| < \delta$, $F(\hat{x} + x) = F(\hat{x}) + y$, $\|x\| \leq K\|y\|$.

PROOF.– We prove the Lagrange theorem by contradiction. Suppose that the stationarity condition

$$\sum_{i=0}^{m} \lambda_i f_i'(\hat{x}) = 0$$

does not hold, and vectors $f_i'(\hat{x})$, $i = 0, 1, \ldots, m$, are linearly independent, which means that the rank of the matrix

$$A = \left\{\frac{\partial f_i(\hat{x})}{\partial x_j}\right\}_{j=\overline{1,n}}^{i=\overline{0,m}}$$

is equal to $m+1$. Therefore, there exists a submatrix of the matrix A of size $(m+1)\times(m+1)$ whose determinant is not zero. Let this be a matrix composed of the first $m+1$

columns of the matrix A. We construct the function $F\colon \mathbb{R}^{m+1} \to \mathbb{R}^{m+1}$ with the help of functions $f_k(x)$, $k = 0, \ldots, m$. Let

$$F_1(x_1, \ldots, x_{m+1}) = f_0(x_1, \ldots, x_{m+1}, \hat{x}_{m+2}, \ldots, \hat{x}_n) - f_0(\hat{x}_1, \ldots, \hat{x}_n),$$

$$F_k(x_1, \ldots, x_{m+1}) = f_{k-1}(x_1, \ldots, x_{m+1}, \hat{x}_{m+2}, \ldots, \hat{x}_n), k = 2, \ldots, m+1.$$

Here $x_1, \ldots, x_{m+1}$ are variables, and $\hat{x}_{m+2}, \ldots, \hat{x}_n$ are fixed quantities. If $\hat{x} = (\hat{x}_1, \ldots, \hat{x}_n)$ is a solution of the constrained optimization problem, then $F(\hat{x}) = 0$. The functions $F_k(x_1, \ldots, x_{m+1})$, $k = 1, \ldots, m+1$ satisfy conditions of the inverse function theorem. Take $y = (\varepsilon, 0, \ldots, 0)$. For sufficiently small values of ε, there exists a vector $\bar{x}(\varepsilon) = (x_1(\varepsilon), \ldots, x_{m+1}(\varepsilon))$ such that

$$F_1(\bar{x}(\varepsilon)) = \varepsilon, \quad F_k(\bar{x}(\varepsilon)) = 0, \quad k = \overline{2, m+1},$$

that is

$$f_0(x(\varepsilon)) - f_0(\hat{x}) = \varepsilon, \quad f_k(x(\varepsilon)) = 0, \quad k = \overline{1, m},$$

where $x(\varepsilon) = (x_1(\varepsilon), \ldots, x_{m+1}(\varepsilon), \hat{x}_{m+2}, \ldots, \hat{x}_n)$ and $\|x(\varepsilon) - \hat{x}\| < K|\varepsilon|$. This contradicts the fact that $\hat{x}$ is a solution to the constrained optimization problem [1.2], since both for positive and negative values of ε there are vectors close to $\hat{x}$ on which the function $f_0(x(\varepsilon))$ takes values of both smaller and larger $f_0(\hat{x})$. □

Thus, to determine $m+n+1$ unknown $\lambda_0, \lambda_1, \ldots, \lambda_m, \hat{x}_1, \ldots, \hat{x}_n$, we have $n+m$ equations

$$f_1(\hat{x}) = \cdots = f_m(\hat{x}) = 0,$$

$$\frac{\partial}{\partial x_j}\left(\sum_{k=0}^{m} \lambda_k f_k(\hat{x})\right) = 0, \quad j = 1, \ldots, n.$$

Note that the Lagrange multipliers are determined up to proportionality. If it is known that $\lambda_0 \neq 0$, then we can choose $\lambda_0 = 1$. Then the number of equations and the number of unknowns are the same.

Linear independence of the vectors of derivatives $f_1'(\hat{x}), \ldots, f_m'(\hat{x})$ is the regularity condition that guarantees that $\lambda_0 \neq 0$. However, verification of this condition is more complicated than direct verification of the fact that λ_0 cannot be equal to zero.

Since the time of Lagrange, almost a century ago, the method of Lagrange multipliers has been used with $\lambda_0 = 1$, despite the fact that in the general case it is wrong.

As in the case of the unconstrained optimization problem, the stationary points of the constrained optimization problem need not be its solution. There also exist the necessary and sufficient conditions for optimality in terms of the second derivatives. Denote by

$$L''_{xx}(x, \lambda, \lambda_0) = \left\{ \frac{\partial^2 L(x, \lambda, \lambda_0)}{\partial x_k \partial x_j} \right\}_{j=1,\dots,n}^{k=1,\dots,n}$$

the matrix of the second-order partial derivatives of the Lagrange function $L(x, \lambda, \lambda_0)$.

THEOREM 1.17.– Let the functions $f_i(x), i = 0, 1, \dots, m$ be two times differentiable at a point $\hat{x}$ and continuously differentiable in some neighborhood U of the point $\hat{x}$, and let the gradients $f_i'(\hat{x}), i = 1, \dots, m$, be linearly independent. If $\hat{x}$ is the point of local minimum of the problem [1.2], then

$$\langle L''_{xx}(\hat{x}, \lambda_0, \lambda)h, h\rangle \geq 0$$

for all λ, λ_0, satisfying the condition

$$L'_x(\hat{x}, \lambda, \lambda_0) = 0,$$

and all $h \in \mathbb{R}^n$ such that

$$\langle f_i'(\hat{x}), h\rangle = 0, \quad i = 1, \dots, m.$$

THEOREM 1.18.– Let the functions $f_i(x), i = 0, 1, \dots, m$, be two times differentiable at a point $\hat{x} \in \mathbb{R}^n$, satisfying the conditions

$$f_i(\hat{x}) = 0, \quad i = 1, \dots, m.$$

Assume that for some λ, λ_0, the condition holds

$$L'_x(\hat{x}, \lambda, \lambda_0) = 0,$$

and, in addition,

$$\langle L''_{xx}(\hat{x}, \lambda, \lambda_0)h, h\rangle > 0$$

for all non-zero $h \in \mathbb{R}^n$ satisfying the condition

$$\langle f_i'(\hat{x}), h\rangle = 0, \quad i = 1, \dots, m.$$

Then $\hat{x}$ is the point of local minimum of the problem [1.2].

We can formulate the following *rule of indeterminate Lagrange multipliers* for solving constrained optimization problems with equality constraints:

1) Write the Lagrange function

$$L(x, \lambda, \lambda_0) = \sum_{k=0}^{m} \lambda_k f_k(x).$$

2) Write the necessary conditions for the extremum of the function L, that is:

$$\frac{\partial}{\partial x_j} L(x, \lambda, \lambda_0) = 0, \quad j = 1, \ldots, n.$$

3) Find the stationary points, that is, the solutions of these equations, provided that not all Lagrange multipliers $\lambda_0, \lambda_1, \ldots, \lambda_m$ are zero.

4) Find a solution to the problem among the stationary points or prove that the problem has no solutions.

1.4.2. *Problems with equality and inequality constraints*

Let $f_i\colon \mathbb{R}^n \to \mathbb{R}$ be differentiable functions of n real variables. The *constrained optimization problem with equality and inequality constraints* is the following problem:

$$f_0(x) \to \inf, \qquad [1.3]$$

$$f_i(x) \leq 0, \quad i = 1, \ldots, m;$$

$$f_{m+k}(x) = 0, \quad k = 1, \ldots, s.$$

We will formulate the necessary conditions for solution of the problem [1.3].

THEOREM 1.19.– (*Theorem on indeterminate Lagrange multipliers*) Let $\hat{x}$ be a point of local solution of the problem [1.3], and let the functions $f_i(x)$, $i = 0, \ldots, m+s$, be continuously differentiable in some neighborhood U of the point $\hat{x}$. Then there will exist the Lagrange multipliers $\lambda_0, \lambda_1, \ldots, \lambda_{m+s}$, not all equal to zero, such that for the Lagrange function

$$L(x, \lambda_0, \ldots, \lambda_{m+s}) = \sum_{i=0}^{m+s} \lambda_i f_i(x)$$

the following conditions hold true:

– stationarity condition with respect to x

$$L_x(\hat{x}, \lambda) = 0 \Longleftrightarrow \frac{\partial L(\hat{x}, \lambda)}{\partial x_j} = 0, \quad j = 1, \ldots, n;$$

– complementary slackness condition

$$\lambda_i f_i(\hat{x}) = 0, \quad i = 1, \ldots, m;$$

– non-negativity condition

$$\lambda_i \geq 0, \quad i = 0, \ldots, m.$$

Consequently, the *rule of indeterminate Lagrange multipliers* for solving constrained optimization problems with equality and inequality constraints is as follows:

1) Write the Lagrange function

$$L(x, \lambda) = \sum_{i=0}^{m+s} \lambda_i f_i(x).$$

2) Write the necessary conditions:

– stationarity condition

$$\frac{\partial L(x, \lambda)}{\partial x_j} = 0, \quad j = 1, \ldots, n;$$

– complementary slackness condition

$$\lambda_i f_i(\hat{x}) = 0, \quad i = 1, \ldots, m;$$

– non-negativity condition

$$\lambda_i \geq 0, \quad i = 0, \ldots, m;$$

3) Find the critical points, that is, all admissible points that satisfy the necessary conditions with the Lagrange multiplier $\lambda_0 = 0$ and $\lambda_0 \neq 0$.

4) Find a solution to the problem among the stationary points or prove that the problem has no solutions.

REMARK 1.1.– Using the rule of indeterminate Lagrange multipliers for solving constrained optimization problems with equality constraints, one can choose the number λ_0 as both positive and negative. For constrained optimization problems with equality and inequality constraints, the sign of λ_0 is significant.

EXAMPLE 1.2.– Solve the constrained optimization problem

$$x_1 \to \inf, \quad x_1^2 + x_2^2 = 0.$$

The only obvious solution to this problem is the point $\hat{x} = (0, 0)$. We solve the problem by the Lagrange method.

1) Write the Lagrange function $L = \lambda_0 x_1 + \lambda(x_1^2 + x_2^2)$.

2) Write the stationary equations

$$L_{x_1} = 0 \Longleftrightarrow 2\lambda x_1 + \lambda_0 = 0, \quad L_{x_2} = 0 \Longleftrightarrow 2\lambda x_2 = 0.$$

3) If $\lambda_0 = 1$, we get the equations

$$2\lambda x_1 + 1 = 0, \quad 2\lambda x_2 = 0.$$

The first equation is incompatible with the condition $x_1^2 + x_2^2 = 0$. Therefore, the system of equations

$$2\lambda x_1 + 1 = 0,$$

$$2\lambda x_2 = 0,$$

$$x_1^2 + x_2^2 = 0$$

has no solutions.

4) If $\lambda_0 = 0$, then $x_1 = 0$, $x_2 = 0$ is a solution of the system of equations.

Answer. $(0, 0) \in \text{absmin}$.

Example 1.2 shows that by applying the rule of indeterminate Lagrange multipliers, it is not always possible to take $\lambda_0 = 1$.

EXAMPLE 1.3.– Solve the constrained optimization problem

$$\frac{1}{2}ax_1^2 + \frac{1}{2}bx_2^2 \to \min, \quad x_1^3 + x_2^3 = 1,$$

where $a > 0$ and $b > 0$ are given numbers.

Solution.

1) Write out the (regular) Lagrange function (as indicated in theorem 1.15 the regularity condition is satisfied here):

$$L(x_1, x_2, \lambda) = \frac{1}{2}ax_1^2 + \frac{1}{2}bx_2^2 + \lambda(x_1^3 + x_2^3 - 1).$$

2) Since

$$L'_{x_1}(x_1, x_2, \lambda) = ax_1 + 3\lambda x_1^2, \quad L'_{x_2}(x_1, x_2, \lambda) = bx_2 + 3\lambda x_2^2,$$

the system of equations for determination of stationary points has the form:

$$ax_1 + 3\lambda x_1^2 = 0, \quad bx_2 + 3\lambda x_2^2 = 0 \quad x_1^3 + x_2^3 = 1.$$

This system of equations has three solutions:

$$\left(0, 1, -\frac{b}{3}\right), \left(1, 0, -\frac{a}{3}\right),$$

$$\left(\frac{a}{(a^3+b^3)^{1/3}}, \frac{b}{(a^3+b^3)^{1/3}}, -\frac{(a^3+b^3)^{1/3}}{3}\right)$$

3) Next we have

$$L''_{xx}(x_1, x_2, \lambda) = \begin{bmatrix} a + 6\lambda x_1 & 0 \\ 0 & b + 6\lambda x_2 \end{bmatrix}.$$

For the three solutions found, this matrix takes the form accordingly

$$A_1 = \begin{bmatrix} a & 0 \\ 0 & -b \end{bmatrix}, \quad A_2 = \begin{bmatrix} -a & 0 \\ 0 & b \end{bmatrix}, \quad A_3 = \begin{bmatrix} -a & 0 \\ 0 & -b \end{bmatrix},$$

The condition $\langle f_i'(\hat{x}), h\rangle = 0$, $i = 1, \ldots, m$, here is of the form $3x_1^2 h_1 + 3x_2^2 h_2 = 0$. For the first two solutions, this means that $h_2 = 0$ and $h_1 = 0$, respectively. It is clear from this that matrices A_1 and A_2 satisfy the conditions of theorem 1.17 (although they are not positive definite). Therefore, the points $(0, 1)$, $(1, 0)$ are strict local solutions of the problem. For the matrix, A_3 conditions of theorem 1.17 are not satisfied. That is why the point

$$\left(\frac{a}{(a^3+b^3)^{1/3}}, \frac{b}{(a^3+b^3)^{1/3}}\right)$$

cannot be a solution of the minimization problem. This point is a local solution of the maximization problem of the same function under the same restrictions.

Answer. $\hat{x}_1 = (0, 1) \in \text{locmin}$, $\hat{x}_2 = (1, 0) \in \text{locmin}$,

$$(\hat{x}_3 = (a/(a^3+b^3)^{1/3}, b/(a^3+b^3)^{1/3}) \in \text{locmax})$$

EXAMPLE 1.4.– Solve the constrained optimization problem

$$x_1^2 + x_2^2 + x_3^2 \to \inf;$$

$$2x_1 - x_2 + x_3 \le 5,$$

$$x_1 + x_2 + x_3 = 3.$$

Solution.

1) Write out the Lagrange function

$$L = \lambda_0(x_1^2 + x_2^2 + x_3^2) + \lambda_1(2x_1 - x_2 + x_3 - 5) + \lambda_2(x_1 + x_2 + x_3 - 3).$$

2) Write the necessary conditions:

– stationarity condition

$$L_{x_1} = 0 \Longleftrightarrow 2\lambda_0 x_1 + 2\lambda_1 + \lambda_2 = 0,$$

$$L_{x_2} = 0 \Longleftrightarrow 2\lambda_0 x_2 + \lambda_2 - \lambda_1 = 0,$$

$$L_{x_3} = 0 \Longleftrightarrow 2\lambda_0 x_3 + \lambda_2 + \lambda_1 = 0;$$

– complementary slackness condition

$$\lambda_1(2x_1 - x_2 + x_3 - 5) = 0;$$

– non-negativity condition $\lambda_0 \geq 0$, $\lambda_1 \geq 0$.

3) If $\lambda_0 = 0$, then, in accordance with the condition of stationarity we have $\lambda_1 = 0$, and $\lambda_2 = 0$. So all Lagrange multipliers are zero. This contradicts the conditions of the Lagrange theorem 1.19. Take $\lambda_0 = 1/2$. Let $\lambda_1 \neq 0$. Then, under the complementary slackness condition we have $2x_1 - x_2 + x_3 - 5 = 0$. We express x_1, x_2, x_3 through λ_1, λ_2 and substitute into the equation

$$x_1 + x_2 + x_3 = 3,$$

$$2x_1 - x_2 + x_3 = 5.$$

We get $\lambda_1 = -9/14 < 0$. This contradicts the non-negativity condition. Let $\lambda_1 = 0$, then $x_1 = x_2 = x_3 = 1$ is a critical point.

4) The function $f(x) = x_1^2 + x_2^2 + x_3^2 \to \infty$ as $\|x\| \to \infty$. By the corollary of the Weierstrass theorem, a solution of the problem exists. Since the critical point is unique, the solution of the problem can be only this point.

Answer. $\hat{x} = (1, 1, 1) \in \text{absmin}$, $S_{\min} = 3$.

EXAMPLE 1.5.– An example of the irregular problem. Consider the constrained optimization problem

$$f(x_1, x_2) = x_1 \to \min,$$

$$g_1(x_1, x_2) = -x_1^3 + x_2 \leq 0,$$

$$g_2(x_1, x_2) = -x_1^3 - x_2 \leq 0,$$

$$g_3(x_1, x_2) = x_1^2 + x_2^2 - 1 \leq 0,$$

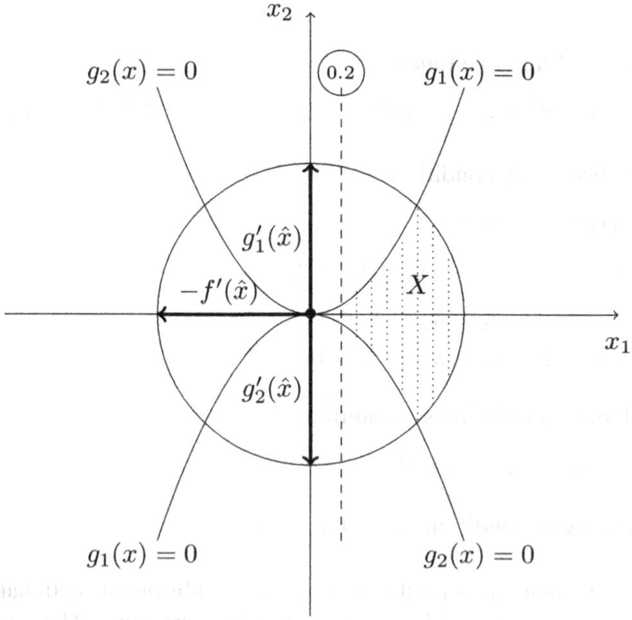

Figure 1.1. *Example 1.5*

Solution. Figure 1.1 depicts the admissible set of the problem and the level line of the objective function. The solution of the problem is the point $\hat{x} = (0, 0)$. Active at this point are the first and the second restrictions. Wherein

$$f'(\hat{x}) = f'(0,0) = (1,0),$$
$$g_1'(\hat{x}) = g_1'(0,0) = (0,1),$$
$$g_2'(\hat{x}) = g_2'(0,0) = (0,-1).$$

The vector $f'(\hat{x}) = f'(0,0) = (1,0)$ cannot be represented as a linear combination of vectors $g_1'(\hat{x}) = g_1'(0,0) = (0,1)$ and $g_2'(\hat{x}) = g_2'(0,0) = (0,-1)$. The relation

$$\lambda_0 f'(\hat{x}) + \lambda_1 g_1'(\hat{x}) + \lambda_2 g_2'(\hat{x}) + \lambda_3 g_3'(\hat{x}) = 0$$

at the point $\hat{x} = (0,0)$ can only be performed with

$$\lambda_0 = 0, \lambda_1 = \lambda, \lambda_2 = -\lambda, \lambda_3 = 0.$$

Gradients $g_1'(\hat{x}) = g_1'(0,0) = (0,1)$ and $g_2'(\hat{x}) = g_2'(0,0) = (0,-1)$ in this case are linearly dependent.

Answer. $\hat{x} = (0,0) \in \text{absmin}, S_{\min} = 0.$

EXAMPLE 1.6.– Solve the convex constrained optimization problem

$$\begin{aligned} f(x_1,x_2) &= x_2 \to \min, \\ g_1(x_1,x_2) &= x_1^2 + x_2^2 - 1 \le 0, \\ g_2(x_1,x_2) &= -x_1 + x_2^2 \le 0, \\ g_3(x_1,x_2) &= x_1 + x_2 \ge 0. \end{aligned}$$

Solution. Slater's condition is fulfilled. Therefore, we write the regular Lagrange function:

$$L(x,y) = x_2 + \lambda_1(x_1^2 + x_2^2 - 1) + \lambda_2(-x_1 + x_2^2) + \lambda_3(-x_1 - x_2).$$

The system for finding stationary points in this case ($s = 0$, $n = 2$, $k = m = 3$) can be written in the form

$$2\lambda_1 x_1 - \lambda_2 - \lambda_3 = 0, \quad 1 + 2\lambda_1 x_2 + 2\lambda_2 x_2 - \lambda_3 = 0,$$

$$\begin{aligned} &\lambda_1 \ge 0, \quad x_1^2 + x_2^2 - 1 \le 0, \quad \lambda_1(x_1^2 + x_2^2 - 1) = 0, \\ &\lambda_2 \ge 0, \quad -x_1 + x_2^2 \le 0, \quad \lambda_2(-x_1 + x_2^2) = 0, \\ &\lambda_3 \ge 0, \quad x_1 + x_2 \ge 0, \quad \lambda_3(x_1 + x_2) = 0. \end{aligned}$$

At point $x = (\sqrt{2}/2, -\sqrt{2}/2)$, the first and third restrictions are active, and the second is passive. Therefore, $\lambda_2 = 0$. As a result, we obtain a system for determining λ_1 and λ_3:

$$\sqrt{2}\lambda_1 - \lambda_3 = 0, \quad 1 - \sqrt{2}\lambda_1 - \lambda_3 = 0, \quad \lambda_1 \ge 0, \quad \lambda_3 \ge 0.$$

System solution $\lambda_1 = \sqrt{2}/4, \lambda_3 = 1/2$. The point $x = (\sqrt{2}/2, -\sqrt{2}/2)$ is a solution of the problem. Make sure that there are no other stationary points and, therefore, no solution to the problem (see Figure 1.2).

Answer. $\hat{x} = (\sqrt{2}/2, -\sqrt{2}/2) \in \text{absmin}, S_{\min} = -\sqrt{2}/2.$

EXAMPLE 1.7.– Let the numbers $a > 0$, $b > 0$, and let $a < b$. Find points of the local minimum and the local maximum of the function

$$f(x) = \frac{1}{2}ax_1^2 + \frac{1}{2}bx_2^2$$

on the set of solutions of the system

$$x_1^3 + x_2^3 \leq 1, \quad x_1^2 + x_2^2 \geq 1.$$

Solution. We denote this set by X. Let us write the Lagrange function

$$L(x, \lambda_0, \lambda) = \lambda_0 \left(\frac{1}{2} a x_1^2 + \frac{1}{2} b x_2^2 \right) + \lambda_1 \left(x_1^3 + x_2^3 - 1 \right) + \lambda_2 \left(-x_1^2 - x_2^2 + 1 \right).$$

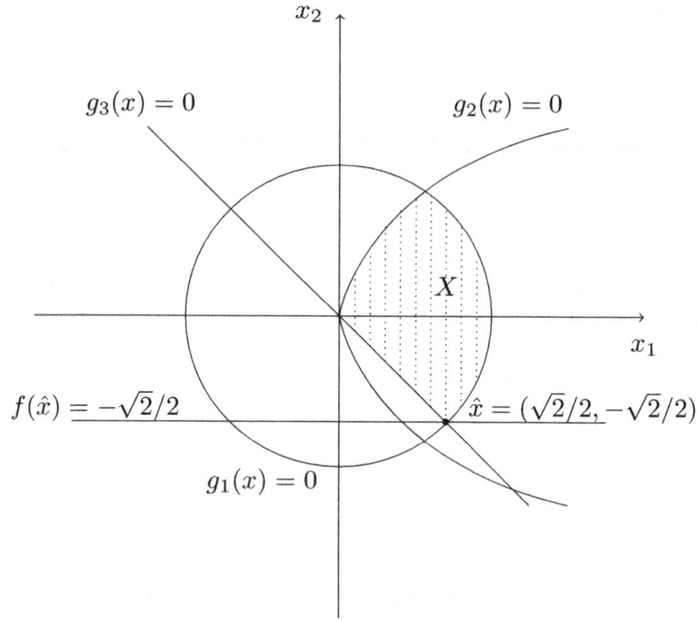

Figure 1.2. *Example 1.6*

The system for determining stationary points has the form

$$a\lambda_0 x_1 + 3\lambda_1 x_1^2 - 2\lambda_2 x_1 = 0. \qquad [1.4]$$

$$b\lambda_0 x_2 + 3\lambda_1 x_2^2 - 2\lambda_2 x_2 = 0. \qquad [1.5]$$

$$\lambda_1 \geq 0, \quad x_1^3 + x_2^3 \leq 1, \quad \lambda_1 \left(x_1^3 + x_2^3 - 1 \right) = 0, \qquad [1.6]$$

$$\lambda_2 \geq 0, \quad x_1^2 + x_2^2 \geq 1, \quad \lambda_2 \left(x_1^2 + x_2^2 - 1 \right) = 0, \qquad [1.7]$$

$$(y_0, y_1, y_2) \neq 0. \qquad [1.8]$$

Let $x_1 = 0$. Then it follows from the system that $x_2 \le 1, x_2^2 \ge 1$. Hence either $x_2 = 1$ or $x_2 \le -1$. In other case, $\lambda_1 = 0$. If $x_2 < -1$, then $\lambda_2 = 0$. But then $\lambda_1 = 0$, which contradicts the conditions of the problem. Now we easily get the first two groups of system solutions.

1) $x_1 = 0, x_2 = 1, b\lambda_0 + 3\lambda_1 - 2\lambda_2 = 0, \lambda_1 \ge 0, \lambda_2 \ge 0, (\lambda_1, \lambda_2) \ne 0$;

2) $x_1 = 0, x_2 = -1, b\lambda_0 - 2\lambda_2 = 0, \lambda_1 = 0, \lambda_2 > 0$.

Similarly, assuming that $x_2 = 0$, we obtain two other groups of solutions:

3) $x_1 = 1, x_2 = 0, a\lambda_0 + 3\lambda_1 - 2\lambda_2 = 0, \lambda_1 \ge 0, \lambda_2 \ge 0, (\lambda_1, \lambda_2) \ne 0$;

4) $x_1 = -1, x_2 = 0, a\lambda_0 - 2\lambda_2 = 0, \lambda_1 = 0, \lambda_2 > 0$.

Assume that $x_1 \ne 0, x_2 \ne 0$. Then equations of the system can be presented in the form

$$a\lambda_0 + 3\lambda_1 x_1 - 2\lambda_2 = 0, \quad b\lambda_0 + 3\lambda_1 x_2 - 2\lambda_2 = 0.$$

If here $\lambda_1 = 0$, then $\lambda_0 = 0$, since $a \ne b$. But then $\lambda_2 = 0$, which contradicts the system of conditions. It remains to be assumed that $\lambda_1 > 0$. Then $x_1^3 + x_2^3 = 1$. Given that $\lambda_1 \ne 0, \lambda_2 \ne 0$, we deduce from this that $x_1^2 + x_2^2 > 1$, and therefore $\lambda_2 = 0$. Now we can easily get another group of solutions of the system:

5) $x_1 = a/\sqrt{3}a^3 + b^3, x_2 = b/\sqrt{3}a^3 + b^3$, $\lambda_0 < 0$, $\lambda_1 = -\lambda_0\sqrt{3}a^3 + b^3/3$, $\lambda_2 = 0$.

Note that in (1) and (3) the multiplier λ_0 can accept both positive and negative values, in (2) and (4) it can accept only positive values and in (5) it can accept only negative values. Therefore, $(0,1)$ and $(1,0)$ are stationary points both in the minimization problem and in the maximization problem, $(0,-1)$ and $(-1,0)$ are stationary points only in the minimization problem, and the point of (5) is the stationary point only in the problem of maximization.

Now we will study the stationary points for optimality. The function f is strongly convex on $\mathbb{R}^2$. Therefore, it reaches the global minimum on any closed set X. We calculate the value of f at the stationary points of the minimization problem:

$$f(0,1) = f(0,-1) = b/2, \quad f(1,0) = f(-1,0) = a/2.$$

Since $a < b$, hence (1.0) and (-1.0) are points of the global minimum of the function f on X.

Let us represent the function f in the form

$$f(x) = \frac{1}{2}a\left(x_1^2 + x_2^2\right) + \frac{1}{2}(b-a)x_2^2.$$

If we move from the points $(0,1)$ and $(0,-1)$, remaining on the circle $x_1^2+x_2^2=1$, and hence in X, then the value of f will decrease. Consequently, these points are not points of the local minimum f on X. At the same time, for any $\varepsilon > 0$, the point (-ε,1) belongs to X and $f(0,1) < f(-\varepsilon,1)$. Therefore, the point $(0,1)$ is not a point of the local maximum f on X. Consequently, the stationary points $(0,1)$ and $(0,-1)$ are not solutions of the problem.

We now consider the matrix of the second derivatives of the Lagrangian function:

$$L''_{xx} = \begin{bmatrix} a\lambda_0 + 6\lambda_1 x_1 - \lambda_2 & 0 \\ 0 & b\lambda_0 + 6\lambda_1 x_1 - \lambda_2 \end{bmatrix}.$$

For values with (5), this matrix is as follows:

$$L''_{xx} = \begin{bmatrix} -a\lambda_0 & 0 \\ 0 & -b\lambda_0 \end{bmatrix}.$$

Since $\lambda_0 < 0$, this matrix is positive definite. Sufficient conditions for the minimum are fulfilled. Consequently, $(a/\sqrt{3}a^3+b^3, b/\sqrt{3}a^3+b^3)$ is the point of the strict local minimum of f on X.

Answer. $\hat{x} = (1.0) \in \text{absmin}$, $\hat{x} = (-1.0) \in \text{absmin}$, $\hat{x} = (a/\sqrt[3]{a^3+b^3}, b/\sqrt[3]{a^3+b^3}) \in \text{absmax}$. $\triangle$

1.5. Exercises

Let us solve the following optimization problems.

1) $f(x,y) = x^4 + y^4 - 4xy \to \text{extr}$.

2) $f(x,y) = ae^{-x} + be^{-y} + \ln(e^x + e^y) \to \text{extr}$.

3) $f(x,y) = (x+y)(x-a)(y-b) \to \text{extr}$.

4) $f(x,y) = x^2 - 2xy^2 + y^4 - y^5 \to \text{extr}$.

5) $f(x,y) = x + y + 4\sin(x)\sin(y) \to \text{extr}$.

6) $f(x,y) = xe^x - (1+e^x)\cos(y) \to \text{extr}$.

7) $f(x,y) = (x^2+y^2)e^{-(x^2+y^2)} \to \text{extr}$.

8) $f(x,y) = xy\ln(x^2+y^2) \to \text{extr}$.

9) $f(x,y) = x - 2y + \ln(\sqrt{x^2+y^2}) + 3\text{arctg}(y/x) \to \text{extr}$.

10) $f(x,y) = \sin(x)\sin(y)\sin(x+y) \to \text{extr}, 0 \le x \le \pi, 0 \le y \le \pi$.

11) $f(x,y) = \sin(x) + \cos(y) + \cos(x-y) \to \text{extr}, 0 \le x \le \pi/2, 0 \le y \le \pi/2$.

12) $f(x,y) = x^2 + xy + y^2 - 4\ln(x) - 10\ln(y) \to \text{extr}$.

13) $f(x,y) = (5x + 7y - 25)e^{-(x^2+y^2+xy)} \to \text{extr}$.

14) $f(x,y) = e^{x^2-y}(5 - 2x + y) \to \text{extr}$.

15) $f(x,y) = e^{2x+3y}(8x^2 - 6xy + 3y^2) \to \text{extr}$.

16) $f(x,y) = 1 - \sqrt{x^2 + y^2} \to \text{extr}$.

17) $f(x,y) = (ax + by + c)/\sqrt{x^2 + y^2 + 1} \to \text{extr}$.

18) $f(x,y) = xy\sqrt{1 - x^2/a^2 - y^2/b^2} \to \text{extr}$.

19) $f(x,y) = 2x^4 + y^4 - x^2 - 2y^2 \to \text{extr}$.

20) $f(x,y) = x^2 - xy + y^2 - 2x + y \to \text{extr}$.

21) $f(x,y) = xy + 50/x + 20/y \to \text{extr}$.

22) $f(x,y) = x^2 - y^2 - 4x + 6y \to \text{extr}$.

23) $f(x,y) = 5x^2 + 4xy + y^2 - 16x - 12y \to \text{extr}$.

24) $f(x,y) = 3x^2 + 4xy + y^2 - 8x - 12y \to \text{extr}$.

25) $f(x,y) = 3xy - x^2y - xy^2 \to \text{extr}$.

26) $f(x,y,z) = x^2 + y^2 + z^2 - xy + x - 2z \to \text{extr}$.

27) $f(x,y,z) = x^2 + 2y^2 + 5z^2 - 2xy - 4yz - 2z \to \text{extr}$.

28) $f(x,y,z) = xy^2z^3(a - x - 2y - 3z) \to \text{extr}, a > 0$.

29) $f(x,y,z) = x^3 + y^2 + z^2 + 12xy + 2z \to \text{extr}, x > 0, y > 0, z > 0$.

30) $f(x,y,z) = x + y^2/4x + z^2/y + 2/z \to \text{extr}$.

31) $f(x,y,z) = x^2 + y^2 + z^2 + 2x + 4y - 6z \to \text{extr}$.

32) $f(x,y) = y \to \text{extr}, \quad x^3 + y^3 - 3xy = 0$.

33) $f(x,y) = x^3 + y^3 \to \text{extr}, \quad ax + by = 1, a > 0, b > 0$.

34) $f(x,y) = x^3/3 + y \to \text{extr}, \quad x^2 + y^2 = a, a > 0$.

35) $f(x,y) = x\sin(y) \to \text{extr}, \quad 3x^2 - 4\cos(y) = 1$.

36) $f(x,y) = x/a + y/b \to \text{extr}, \quad x^2 + y^2 = 1$.

37) $f(x,y) = x^2 + y^2 \to \text{extr}, \quad x/a + y/b = 1$.

38) $f(x,y) = Ax^2 + 2Bxy + Cy^2 \to \text{extr}, \quad x^2 + y^2 = 1$.

39) $f(x,y) = x^2 + 12xy + 2y^2 \to \text{extr}, \quad 4x^2 + y^2 = 25$.

40) $f(x,y) = \cos^2(x) + \cos^2(y) \to \text{extr}, \quad x - y = \pi/4$.

41) $f(x,y) = x/2 + y/3 \to \text{extr}, \quad x^2 + y^2 = 1.$

42) $f(x,y) = x^2 + y^2 \to \text{extr}, \quad 3x + 4y = 1.$

43) $f(x,y) = e^{xy} \to \text{extr}, \quad x + y = 1.$

44) $f(x,y) = 5x^2 + 4xy + y^2 \to \text{extr}, \quad x + y = 1.$

45) $f(x,y) = 3x^2 + 4xy + y^2 \to \text{extr}, \quad x + y = 1.$

46) $f(x,y,z) = xy^2z^3 \to \text{extr}, \quad x + y + z = 1.$

47) $f(x,y,z) = xyz \to \text{extr}, \quad x^2 + y^2 + z^2 = 1, x + y + z = 0.$

48) $f(x,y,z) = a^2x^2 + b^2y^2 + c^2z^2 - (ax^2 + by^2 + cz^2)^2 \to \text{extr}, x^2 + y^2 + z^2 = 1, a > b > c > 0.$

49) $f(x,y,z) = x + y + z^2 + 2(xy + yz + zx) \to \text{extr}, x^2 + y^2 + z = 1.$

50) $f(x,y,z) = x - 2y + 2z \to \text{extr}, \quad x^2 + y^2 + z^2 = 1.$

51) $f(x,y,z) = x^m y^n z^p \to \text{extr}, x + y + z = a, m > 0, n > 0, p > 0, a > 0.$

52) $f(x,y,z) = x^2 + y^2 + z^2 \to \text{extr}, \ x^2/a^2 + y^2/b^2 + z^2/c^2 = 1, a > b > c > 0.$

53) $f(x,y,z) = xy^2z^3 \to \text{extr}, \quad x + 2y + 3z = a, x > 0, y > 0, z > 0, a > 0.$

54) $f(x,y,z) = xy + yz \to \text{extr}, x^2 + y^2 = 2, y + z = 2, x > 0, y > 0, z > 0.$

55) $f(x,y,z) = \sin(x)\sin(y)\sin(z) \to \text{extr}, x + y + z = \pi/2.$

56) $f(x,y) = e^{x-y} - x - y \to \text{extr}, x + y \leq 1, x \geq 0, y \geq 0.$

57) $f(x,y) = x^2 + y^2 - 2x - 4y \to \text{extr}, 2x + 3y - 6 \leq 0, x + 4y - 5 \leq 0.$

58) $f(x,y) = 2xy - x^2 - 2y^2 \to \text{extr}, x - y + 1 \geq 0, 2x + 3y + 6 \leq 0.$

59) $f(x,y) = x^2 + y^2 \to \text{extr}, -5x + 4y \leq 0, -x + 4y + 3 \leq 0.$

60) $f(x,y) = x^2 + y^2 - 2x \to \text{extr}, x - 2y + 2 \leq 0, 2x - y \geq 0.$

61) $f(x,y,z) = xyz \to \text{extr}, \quad x^2 + y^2 + z^2 \leq 1.$

62) $f(x,y,z) = 2x^2 + 2x + 4y - 3z \to \text{extr}, 8x - 3y + 3z \leq 40, -2x + y - z = -3, y \geq 0.$

63) $f(x,y,z) = x^2 + 4y^2 + z^2 \to \text{extr}, x + y + z \leq 12, x \geq 0, y \geq 0, z \geq 0.$

64) $f(x,y,z) = 3y^2 - 11x - 3y - z \to \text{extr}, x - 7y + 3z + 7 \leq 0, 5x + 2y - z \leq 2, z \geq 0.$

65) $f(x,y,z) = xz - 2y \to \text{extr}, 2x - y - 3z \leq 10, 3x + 2y + z = 6, y \geq 0.$

66) $f(x,y,z) = -4x - y + z^2 \to \text{extr}, x^2 + y^2 + xz - 1 \leq 0, x^2 + y^2 - 2y \leq 0, 5 - x + y + z \leq 0, x \geq 0, y \geq 0, z \geq 0.$

67) $\prod_{j=1}^{n} x_j^{\alpha_j} \to \max, \sum_{j=1}^{n} a_j x_j^{\beta_j} = b,\ b > 0, x_j \geq 0, \alpha_j > 0, \beta_j > 0, a_j > 0, j = 1, 2, \ldots, n.$

68) $\sum_{j=1}^{n} c_j x_j^{\alpha_j} \to \min, \quad \prod_{j=1}^{n} x_j^{\beta_j} = b,\ b > 0, x_j \geq 0, c_j > 0, \alpha_j > 0, \beta_j > 0, j = 1, 2, \ldots, n.$

69) $\sum_{j=1}^{n} \frac{c_j}{x_j^{\alpha_j}} \to \min, \quad \prod_{j=1}^{n} x_j^{\beta_j} = b,\ b > 0, x_j > 0, c_j > 0, \alpha_j > 0, \beta_j > 0, j = 1, 2, \ldots, n.$

70) $\sum_{j=1}^{n} \frac{c_j}{x_j^{\alpha}} \to \min, \quad \sum_{j=1}^{n} a_j x_j = b,\ b > 0, \alpha > 0, x_j > 0, c_j > 0, j = 1, 2, \ldots, n.$

71) $\sum_{j=1}^{n} c_j x_j^{\alpha} \to \max, \quad \sum_{j=1}^{n} a_j x_j = b,\ b > 0, 0 < \alpha < 1, x_j > 0, c_j > 0, j = 1, 2, \ldots, n.$

72) $\sum_{j=1}^{n} c_j x_j^{\alpha} \to \min, \quad \sum_{j=1}^{n} a_j x_j = b,\ c_j > 0, a = (a_1, \ldots, a_n) \neq 0, \alpha = 2m, m \in N.$

73) $\sum_{j=1}^{n} c_j |x_j|^{\alpha} \to \min, \quad \sum_{j=1}^{n} a_j x_j = b,\ c_j > 0, a = (a_1, \ldots, a_n) \neq 0, \alpha > 1, b > 0.$

74) $\sum_{j=1}^{n} c_j x_j \to \max(\min), \quad \sum_{j=1}^{n} a_j x_j^{\alpha} = b,\ b > 0, a_j > 0, c = (c_1, \ldots, c_n) \neq 0, \alpha = 2m, m \in N.$

75) $\sum_{j=1}^{n} c_j x_j \to \max(\min), \quad \sum_{j=1}^{n} a_j |x_j|^{\alpha} = b,\ b > 0, a_j > 0, c = (c_1, \ldots, c_n) \neq 0, \alpha > 1.$

76) $\sum_{j=1}^{n} |c_j + x_j|^{\alpha} \to \max(\min), \quad \sum_{j=1}^{n} |x_j|^{\alpha} = b, b > 0, c = (c_1, \ldots, c_n) \neq 0, \alpha > 1.$

77) Divide the number 8 into two parts so that the product of their product on the difference is maximal (Niccolo Tartaglia problem).

78) Determine the rectangular triangle of the largest area, provided that the sum of the lengths of its legs is equal to a given number (Fermat's problem).

79) On the BC side of a triangle ABC, define a point E such that the parallelogram $ADEK$, whose points D and K lie on sides AB and AC, respectively, has the largest area (Euclid problem).

80) On a given face of a tetrahedron, take a point through which planes parallel to three other faces are drawn. Choose a point so that the volume of the parallelepiped is maximal (generalized Euclid problem).

81) Determine a polynomial of the second-degree $t^2 + x_1 t + x_2$ such that the integral

$$\int_{-1}^{1} (t^2 + x_1 t + x_2)^2 \, dt$$

gets the smallest value (Legendre problem for polynomial of the second degree).

82) Determine a polynomial of the third degree $t^3 + x_1t^2 + x_2t + x_3$ such that the integral

$$\int_{-1}^{1} (t^3 + x_1t^2 + x_2t + x_3)^2 \, dt$$

gets the smallest value (Legendre problem for polynomial of the third degree).

83) Among all discrete random variables that take n values, determine a random variable with the largest entropy. The entropy of the sequence of positive numbers $p_1, \ldots, p_n$, such that $\sum_{k=1}^{n} p_k = 1$, is the number

$$H = -\sum_{i=1}^{n} p_i \ln(p_i).$$

84) Insert a rectangle of maximum area into a circle.

85) Insert a triangle of maximum area into a circle.

86) Insert a cylinder with maximum volume into a ball (Kepler's problem).

87) Insert a cone with maximum volume into a ball.

88) Among cones inscribed in a ball, determine a cone with the maximum area of the lateral surface.

89) Insert in a sphere from the space $\mathbb{R}^n$ a rectangular parallelepiped with the largest volume.

90) Insert a tetrahedron with the largest volume into a ball.

91) Among triangles with a given perimeter determine a triangle of the largest area.

92) Among all n-angles of a given perimeter determine an n-cube of the largest area (Zeno's problem).

93) Insert n-angles of the maximum area in a circle.

94) On the diameter AB of a circle of the unit radius, a point E is taken through which a chord CD is drawn. Determine a position of the chord in which the square of the quadrilateral $ABCD$ is maximal.

95) Determine in a triangle such a point that the sum of the ratio of lengths of sides and distances from the point to relevant sides is minimal.

96) Insert into a circle a triangle with the largest sum of squares of sides.

97) Through a given point inside a corner, draw a segment with ends on the sides of the corner so that the area of the formed triangle is minimal.

98) Through a point inside a corner draw a section with ends on the sides of the corner so that the perimeter of the formed triangle is minimal.

99) Determine a quadrilateral with given sides of the largest area.

100) Among segments of a ball having a given area of the lateral surface, find the segment with the largest volume (Archimedes' problem).

101) Determine a point C on a line such that the sum of the distances from the point C to the given points A and B is minimal.

102) Among all tetrahedra with a given base and height, find a tetrahedron with the smallest lateral surface.

103) Among all tetrahedra with a given base and area of the lateral surface, find a tetrahedron with the largest volume.

104) Among all tetrahedra having a given area of the surface, find a tetrahedron that has the largest volume.

105) Three points x_1, x_2, x_3 are given on a plane. Determine a point x_0 such that the sum of the squares of distances from the point x_0 to the points x_1, x_2, x_3 is the smallest.

106) In the space $\mathbb{R}^n$ there are given N points $x_1, \ldots, x_N$ and N positive numbers $m_1, \ldots, m_N$. Determine a point x_0, such that the sum with the coefficients m_i of the squares of distances from the point x_0 to $X_1, \ldots, x_N$ is the smallest.

107) Solve the previous problem, provided that the point x_0 lies on the sphere of unit radius.

108) Solve the previous problem, provided that the point x_0 belongs to the ball of unit radius.

109) Find the distance from a point to the ellipse. How many normals can be drawn from a point to the ellipse (Apollonius's problem)?

110) Find the distance from a point x_0 to a parabola.

111) Find the distance from a point x_0 to a hyperbole.

112) Find the distance from a point x_0 in the space $\mathbb{R}^n$ to the hyperplane $H = \{x \in \mathbb{R}^n | \langle a, x \rangle = \beta\}$.

113) Find the distance from a point x_0 to the hyperplane in a Hilbert space.

114) Find the distance from a point x_0 in the space $\mathbb{R}^n$ to a line.

115) Find the minimum of a linear function in the space $\mathbb{R}^n$ on a unit ball.

116) In the ellipse $x^2/a^2 + y^2/b^2 = 1$ insert a rectangle of the largest area with sides parallel to the coordinate axes.

117) In the ellipsoid $x^2/a^2 + y^2/b^2 + z^2/c^2 = 1$ insert a rectangular parallelepiped of the largest volume with edges parallel to the axes of coordinates.

118) Prove the inequality between the power averages

$$\left(\frac{1}{n}\sum_{i=1}^{n}|x_i|^p\right)^{\frac{1}{p}} \leq \left(\frac{1}{n}\sum_{i=1}^{n}|x_i|^q\right)^{\frac{1}{q}}, -\infty < p \leq q \leq \infty, p \neq 0, q \neq 0,$$

solving the problem

$$\sum_{i=1}^{n}|x_i|^p \to \max, \sum_{i=1}^{n}|x_i|^q = a^q, 1 < p < q, a > 0.$$

119) Prove the inequality

$$\left(\sum_{i=1}^{n}|x_i|^p\right)^{\frac{1}{p}} \leq \left(\sum_{i=1}^{n}|x_i|^q\right)^{\frac{1}{q}}, \quad 0 < q \leq p \leq \infty.$$

120) Prove the inequality

$$\left(\prod_{i=1}^{n}x_i\right)^{\frac{1}{n}} \leq \frac{1}{n}\sum_{i=1}^{n}x_i \quad \text{for all} \quad x_i \geq 0,\ i = 1, \ldots, n.$$

121) Prove the Hölder inequality

$$\left|\sum_{i=1}^{n}x_i y_i\right| \leq \left(\sum_{i=1}^{n}|x_i|^p\right)^{\frac{1}{p}}\left(\sum_{i=1}^{n}|y_i|^q\right)^{\frac{1}{q}}, \quad \frac{1}{p}+\frac{1}{q} = 1, p > 1, q > 1.$$

Make sure that for $y = (y_1, \ldots, y_n) \neq 0$, the equality holds only when $|x_i| = \lambda|y_i|, i = 1, \ldots, n$.

122) Prove the Minkowski inequality

$$\left(\sum_{i=1}^{n}|x_i + y_i|^p\right)^{\frac{1}{p}} \leq \left(\sum_{i=1}^{n}|x_i|^p\right)^{\frac{1}{p}} + \left(\sum_{i=1}^{n}|y_i|^p\right)^{\frac{1}{p}}, p > 1.$$

Make sure that for $y = (y_1, \ldots, y_n) \neq 0$, the equality holds only when $x_i = \lambda y_i, \lambda > 0, i = 1, \ldots, n$.

2

Convex Sets

2.1. Convex sets: basic definitions

We deal with the n-dimensional space $\mathbb{R}^n$ equipped with the scalar product $\langle x, y \rangle$, $x, y \in \mathbb{R}^n$, so it is Euclidean space and also a complete normed space with the norm $\|x\| = \sqrt{\langle x, x \rangle}$.

DEFINITION 2.1.– *The interval that connects points x^1 and x^2 of the n-dimensional space $\mathbb{R}^n$ is the set*

$$[x^1, x^2] = \{x \in \mathbb{R}^n \colon x = \lambda x^1 + (1 - \lambda) x^2,\ 0 \le \lambda \le 1\}.$$

DEFINITION 2.2.– *A non-empty set $X \subset \mathbb{R}^n$ is called convex if, together with any two points, it contains the interval joining these points. We shall assume that the empty set $\emptyset$ is convex by definition.*

Examples of convex sets in the space $\mathbb{R}^1$ are single-point sets, intervals, rays and straight lines.

Examples of convex sets in the space $\mathbb{R}^n$ are the space itself, any subspace, one-point sets, balls, segments and also:

1) the straight line passing through a point $\hat{x}$ in the direction h

$$l_{\hat{x}h} = \{x \in \mathbb{R}^n \colon x = \hat{x} + \alpha h\,, \alpha \in \mathbb{R}\};$$

2) the ray proceeding from a point $\hat{x}$ in the direction h

$$l^+_{\hat{x}h} = \{x \in \mathbb{R}^n \colon x = \hat{x} + \alpha h\,, \alpha \in \mathbb{R}, \alpha \ge 0\};$$

3) the hyperplane with the normal vector p

$$H_{p\beta} = \{x \in \mathbb{R}^n \colon \langle p, x \rangle = \beta\};$$

4) half-spaces generated by this hyperplane

$$H^+_{p\beta} = \{x \in \mathbb{R}^n : \langle p, x\rangle \geq \beta\};$$

$$H^-_{p\beta} = \{x \in \mathbb{R}^n : \langle p, x\rangle \leq \beta\}.$$

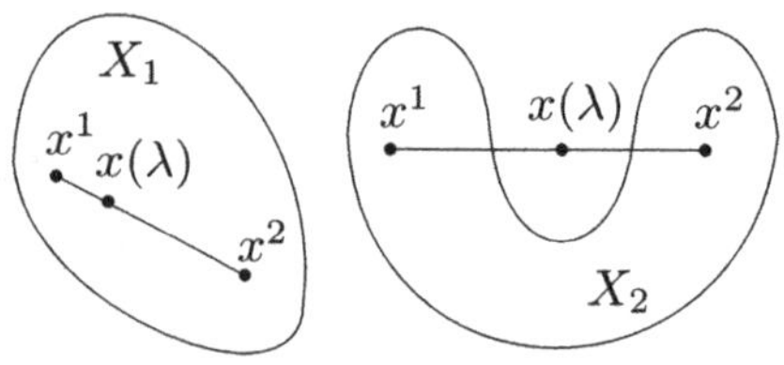

Figure 2.1. *Convex set X_1. Non-convex set X_2*

THEOREM 2.1.– Let I be a set of indices (finite or infinite), and let $X_i, i \in I$, be convex sets. The intersection $X = \cap_{i\in I} X_i$ of convex sets is a convex set.

PROOF.– Let $x^1, x^2 \in X, \lambda \in [0,1]$. By definition of intersection of sets points $x^1, x^2 \in X_i$ for all $i \in I$. Since X_i are convex sets, then $x = \lambda x^1 + (1-\lambda)x^2 \in X_i$. Hence $x \in \cap_{i\in I} X_i = X$. So the set $X = \cap_{i\in I} X_i$ is convex. □

The proof of the following theorem is as simple.

THEOREM 2.2.– Let $X_1, \ldots, X_m$ be convex sets, and let $a_1, \ldots, a_m$ be arbitrary numbers. Then the set

$$\sum_{i=1}^m a_i X_i = \left\{ x \,\middle|\, x = \sum_{i=1}^m a_i x^i,\ x^i \in X_i,\ \ i = 1, \ldots, m \right\},$$

which is called the linear combination of sets $X_1, \ldots, X_m$, is convex.

As a result, the sum and the difference

$$X_1 \pm X_2 = \{x | x = x^1 \pm x^2,\ x^1 \in X_1,\ x^2 \in X_2\}$$

of convex sets X_1, X_2 are convex.

THEOREM 2.3.– Let X_1, X_2 be convex sets. Then the set

$$X_1 \boxplus X_2 = \cup\{\alpha_1 X_1 \cap \alpha_2 X_2 : \alpha_1 \geq 0,\ \alpha_2 \geq 0,\ \alpha_1 + \alpha_2 = 1\};$$

which is called the convolution of sets X_1, X_2, is convex.

THEOREM 2.4.– Let X_1, X_2 be convex sets. Then the set

$$X_1 \text{ conv } \cup X_2 = \text{conv } (X_1 \cup X_2) =$$

$$= \{x \colon x = \alpha_1 x_1 + \alpha_2 x_2, x_1 \in X_1, x_2 \in X_2, \alpha_1 \geq 0, \alpha_2 \geq 0, \alpha_1 + \alpha_2 = 1\};$$

which is called the convex hull of the union of sets X_1, X_2, is convex.

Important subclasses of convex sets form convex cones and affine sets.

DEFINITION 2.3.– *A set $K \in \mathbb{R}^n$ is called:*

a) a cone, if from the fact that $x \in K$ and $\lambda \geq 0$, it follows that $\lambda x \in K$;

b) a convex cone, if from the fact that $x^1, x^2 \in K$ and $\lambda_1 \geq 0$, $\lambda_2 \geq 0$, it follows that $\lambda_1 x^1 + \lambda_2 x^2 \in K$.

A convex cone contains all linear combinations of its elements with positive coefficients.

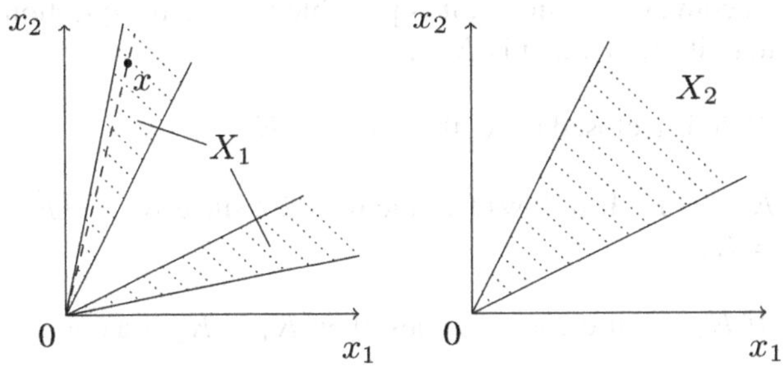

Figure 2.2. X_1 *is a cone.* X_2 *is a convex cone*

DEFINITION 2.4.– *The set of vectors $x^* \in \mathbb{R}^n$ such that $\langle x, x^* \rangle \geq 0 \, \forall x \in K$ is called a conjugate cone to the cone K and denoted by K^*. It can be written as follows:*

$$K^* = \{x^* \in \mathbb{R}^n \colon \langle x, x^* \rangle \geq 0, x \in K\}.$$

Here are some properties of conjugate cones. Obviously, K^* is a convex cone.

LEMMA 2.1.– The cone K^* is a closed cone.

PROOF.– If $x_k^* \in K^*, x_k^* \to x_0^*$, then on passing to the limit in the inequality $\langle x, x_k^* \rangle \geq 0$ we obtain $\langle x, x_0^* \rangle \geq 0 \, \forall x \in K$. Hence $x_0^* \in K^*$. □

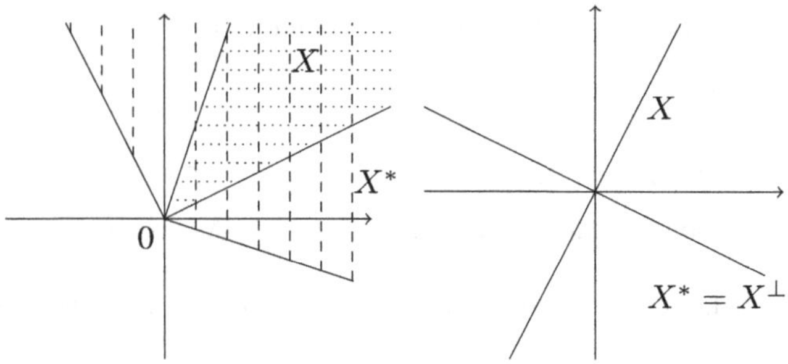

Figure 2.3. *Conjugate cones*

LEMMA 2.2.– Cones K and $\overline{K}$ have the same conjugate cones, i.e. $K^* = (\overline{K})^*$.

LEMMA 2.3.– If K is a closed cone and $\langle x, x^* \rangle \geq 0 \; \forall x^* \in K^*$, then $x \in K$.

Since K^* is a convex cone in $\mathbb{R}^n$, it is possible to raise the question of calculating a cone conjugate to it $(K^*)^*$, that is K^{**}.

LEMMA 2.4.– If K is a closed cone, then $K^{**} = K$.

In general, $K^{**} = \overline{K}$. It follows from lemma 2.2 since $K^* = (\overline{K})^*$ and therefore $K^{**} = (\overline{K})^{**} = \overline{K}$.

LEMMA 2.5.– If K_1, K_2 are convex cones, then $K_1 + K_2$ is a convex cone and

$$(K_1 + K_2)^* = K_1^* \cap K_2^*.$$

The next lemma is very important.

LEMMA 2.6.– For closed cones K_1 and K_2, the following equality holds true:

$$(K_1 \cap K_2)^* = \overline{K_1^* + K_2^*}$$

PROOF.– The proof of the lemma is formally based on previous results:

$$(K_1 \cap K_2)^* = (K_1^{**} \cap K_2^{**})^* = ((K_1^* + K_2^*)^*)^* = (K_1^* + K_2^*)^{**} = \overline{K_1^* + K_2^*}.$$

□

DEFINITION 2.5.– *A set* $X \subset \mathbb{R}^n$ *is called affine if* $\lambda x^1 + (1 - \lambda) x^2 \in X$ *for all* $x^1, x^2 \in X$, $\lambda \in \mathbb{R}$, *that is, if* X *together with its two points* x^1, x^2 *contains a straight line passing through these points.*

Affine sets have a very simple structure: they are shifts of linear subspaces, or sets of solutions of systems of finite number of linear equations, or the intersection of a finite number of hyperplanes.

THEOREM 2.5.– Let X be an affine set in $\mathbb{R}^n$. Then:

1) for arbitrary $x^0 \in X$, the set $L = X - x^0$ is a linear subspace, and L does not depend on the choice x^0 in X;

2) the set X can be written as:

$$X = \{x \in \mathbb{R}^n \,|Ax = b\} = \{x \in \mathbb{R}^n \,|\langle a_i, x\rangle = b_i,\, i = 1, \ldots, m\}, \qquad [2.1]$$

where A is a matrix of dimension $m \times n$ with rows $a_1, \ldots, a_m$, $a_i = (a_{i1}, \ldots, a_{in})$, $b = (b_1, \ldots, b_m) \in \mathbb{R}^m$.

PROOF.– 1) It is clear that L is an affine set and $0 \in L$. Then for arbitrary $x^1,\ x^2 \in L$, $\lambda \in \mathbb{R}$, we have $\lambda x^1 = \lambda x^1 + (1 - \lambda) \cdot 0 \in L$, $x = 0,5x^1 + 0,5x^2 \in L$ and $x^1 + x^2 = 2x \in L$. Consequently, the sum of two elements of L and the product of an element from L by a number belongs to L, that is, L is a linear subspace. Let $L^1 = X - x^1$, where $x^1 \in X$. Take an arbitrary point $x \in L$. Since $x^1 - x^0 \in L$, then $x + x^1 - x^0 \in L$ and $x \in L + x^0 - x^1 = X - x^1 = L^1$. Therefore, $L \subset L^1$. The inclusion of $L^1 \subset L$ is proved similarly. Hence $L^1 = L$.

2) It is known that any linear subspace can be represented as the set of solutions of a system of homogeneous linear equations. Let L in (1) be of the form $L = \{x \in \mathbb{R}^n \mid Ax = 0\}$. Then for $b = Ax^0$, we get [2.1]. □

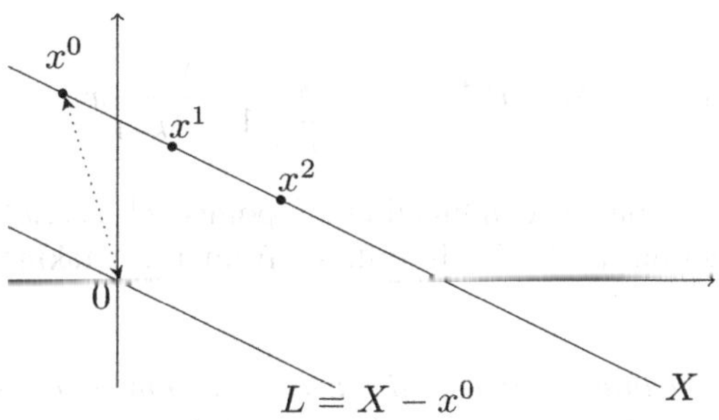

Figure 2.4. *Affine set and linear subspace*

COROLLARY 2.1.– Any affine set $X \subset \mathbb{R}^n$ is closed, moreover, $X = \mathbb{R}^n$ (in [2.1] this case corresponds to $A = 0,\ b = 0$), or $\operatorname{int} X = \emptyset$, that is, X has no interior points.

For affine sets, analogs of theorems 2.1 and 2.2 hold true.

2.2. Combinations of points and hulls of sets

DEFINITION 2.6.– *Let $x^1, \dots, x^m$ be points from $\mathbb{R}^n$. The combination $\sum_{i=1}^{m} \lambda_i x^i$ of points $x^1, \dots, x^m$ is said to be:*

1) convex, if $\lambda_i \geq 0, i = 1, \dots, m,\ \sum_{i=1}^{m} \lambda_i = 1$;

2) conical, if $\lambda_i \geq 0,\ i = 1, \dots, m$;

3) affine, if $\sum_{i=1}^{m} \lambda_i = 1$;

4) linear, if $\lambda_i,\ i = 1, \dots, m$, are real numbers from $\mathbb{R}$.

THEOREM 2.6.– The convex set (convex cone, affine set, linear space) contains all possible convex (conical, affine, linear) combinations of its points.

PROOF.– The statement for convex cones and linear spaces follows immediately from definitions, and the statement for affine sets follows from theorem 2.5. We prove the statement for a convex set X. It is necessary to show that for arbitrary $m = 1, 2, \dots$ from the conditions

$$x = \sum_{i=1}^{m} \lambda_i x^i,\ x^i \in X,\ \lambda_i \geq 0,\ i = 1, \dots, m;\ \sum_{i=1}^{m} \lambda_i = 1, \qquad [2.2]$$

it follows that $x \in X$. We prove the statement by induction on m. If $m = 1$, then the statement is trivial. Assume that the statement is proved for $m = k$, and let [2.2] be written for $m = k + 1$. If $\lambda_{k+1} = 1$, then $\lambda_1 = \dots = \lambda_k = 0$, and, respectively, $x = x^{k+1} \in X$. If $\lambda_{k+1} < 1$, then we can write

$$x = (1 - \lambda_{k+1})\overline{x} + \lambda_{k+1} x^{k+1}, \quad \overline{x} = \sum_{i=1}^{k} \frac{\lambda_i}{1 - \lambda_{k+1}} x^i. \qquad [2.3]$$

The point $\overline{x}$ is a convex combination of points $x^1, \dots, x^k$. Then, from the assumption of induction $\overline{x} \in X$. It follows from [2.3] taking into account the convexity of X, that $x \in X$. □

DEFINITION 2.7.– *The intersection of all convex sets (convex cones, affine sets, linear spaces) from $\mathbb{R}^n$ containing a given set X is called the convex (conic, affine, linear) hull of the set X and it is denoted by conv X (cone X, aff X, Lin X).*

For any set $X \subset \mathbb{R}^n$, its convex hull conv X is non-empty (since X is contained at least in the space $\mathbb{R}^n$, which is a convex set). If the set Y is convex and contains X, then by definition conv $X \subset Y$. In other words, conv X is the smallest set containing X. It is clear that the set X is convex only in the case when conv $X = X$. Similar remarks can be made with respect to convex cones and conical hulls and affine sets and affine hulls.

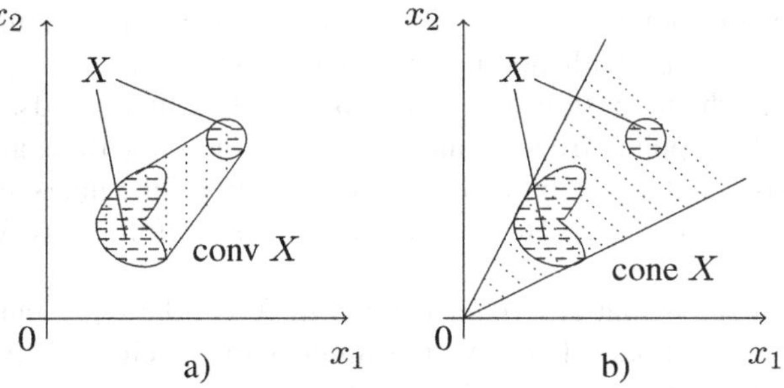

Figure 2.5. *a) Convex hull. b) Conic hull*

DEFINITION 2.8.– *Lin X, by definition, is a linear subspace parallel to the affine hull aff X of the set X.*

Thus $\operatorname{Lin} X = \operatorname{aff} X - x^0$, where x^0 is an arbitrary point from X, or even from $\operatorname{aff} X$, and $\operatorname{Lin} X$ is uniquely determined by theorem 2.5.

LEMMA 2.7.– The linear subspace $\operatorname{Lin} X$ has the following properties:

1) if $x^1, x^2 \in X$, then $x^1 - x^2 \in \operatorname{Lin} X$;

2) if $x \in X$, h^1, $h^2 \in \operatorname{Lin} X$ and α_1, $\alpha_2 \in \mathbb{R}$, then

$$x + \alpha_1 h^1 + \alpha_2 h^2 \in \operatorname{aff} X;$$

3) if $0 \in X$, then $\operatorname{Lin} X = \operatorname{aff} X$.

THEOREM 2.7.– The convex (conical, affine, linear) hull of an arbitrary set X coincides with the set of all convex (conical, affine, linear) combinations of points from X.

PROOF.– Consider the case of a convex hull. For the conical and affine hulls, the proof is similar. Denote by Z the set of all possible convex combinations of points from X. We need to show that $\operatorname{conv} X = Z$. Let us verify that Z is convex. Let $x, y \in Z$ and $\lambda \in [0, 1]$. By the definition of Z, we have:

$$x = \sum_{i=1}^{m} \mu_i x^i, x^i \in X,\ \mu_i \geq 0,\ i = 1, \ldots, m;\ \sum_{i=1}^{m} \mu_i = 1,$$

$$y = \sum_{k=1}^{s} \eta_k y^k, y^k \in X,\ \eta_k \geq 0,\ k = 1, \ldots, s;\ \sum_{k=1}^{s} \eta_k = 1.$$

In this case, the point $z = \lambda x + (1-\lambda)\, y$ is a linear combination of points $x^1, \ldots, x^m$, y^1, ..., y^s with non-negative coefficients $\lambda\mu_1, \ldots, \lambda\mu_m$, $(1-\lambda)\,\eta_1$, ..., $(1-\lambda)\,\eta_s$, which in sum are equal to one. In other words, z is a convex combination of these points from X and $X \subset Z$. At the same time, any convex set Y containing X also contains Z by theorem 2.6. Therefore, the intersection of all such Y, that is, conv X, contains Z. Hence conv $X \supset Z$. Therefore, conv $X = Z$. □

This theorem states that any point from conv X can be represented as a convex combination of some points from X, the number of which, of course, is not very large. It turns out that for $X \subset \mathbb{R}^n$, this number can always be limited to $n+1$. This assertion, known as the Carathéodory theorem, is one of the most important facts in finite-dimensional convex analysis.

THEOREM 2.8.– (*Carathéodory theorem*) In the space $\mathbb{R}^n$, any point from the convex (conic, affine, linear) hull of a set X can be represented as a convex (conic, affine, linear) combination of not more than r points from X. For convex and affine combinations, $r = n+1$. For conic and linear combinations, $r = n$.

PROOF.– We present a proof of the theorem for the convex hull. It is clear that the central place in the theorem is the assertion that $r \leq n+1$. Take a point of the form [2.2] and show that the number of non-zero summands in sum [2.2] can be reduced if $m > n+1$. It is enough to assume that all $\lambda_i > 0$. We take $(n+1)$-dimensional vectors $(x^i, 1), i = 1, \ldots, m$, in which the first n components coincide with corresponding components of the vector x^i and the last component is equal to 1. Since the number of such vectors is $m > n+1$, they are linear dependent. Therefore, there are numbers $\alpha_i, i = 1, \ldots, m$ which are not all equal to zero such that

$$\sum_{i=1}^{m} \alpha_i x^i = 0, \quad \sum_{i=1}^{m} \alpha_i = 0.$$

Among the numbers α_i, there are necessarily positive numbers due to the second relation. Take

$$\varepsilon_0 = \min\left\{\frac{\lambda_i}{\alpha_i} : \alpha_i > 0\,, i = 1, \ldots, m\right\}.$$

The minimum is achieved for some $i = i_0$. Then

$$\bar{\lambda}_i = \lambda_i - \varepsilon_0 \alpha_i \geq 0, \quad i = 1, \ldots, m.$$

This is obvious for $\alpha_i \leq 0$, and for $\alpha_i > 0$ follows from the definition of ε_0. Now from the relationships

$$\sum_{i=1}^{m} \bar{\lambda}_i x^i = \sum_{i=1}^{m} \lambda_i x^i - \varepsilon_0 \sum_{i=1}^{m} \alpha_i x^i = x,$$

$$\sum_{i=1}^{m} \bar{\lambda}_i = \sum_{i=1}^{m} \lambda_i - \varepsilon_0 \sum_{i=1}^{m} \alpha_i = 1$$

it follows that the point x can be given as a convex combination of fewer non-zero terms. We can apply these procedures as long as $m > n + 1$. Hence the statement of the theorem is proved. □

COROLLARY 2.2.– If X is a compact set in $\mathbb{R}^n$, then $\operatorname{conv} X$ is also a compact set.

PROOF.– Consider the Cartesian product $Y = \Lambda \times X \times \cdots \times X$, where X is taken $n+1$ times, and

$$\Lambda = \left\{ \lambda \in \mathbb{R}^{n+1} \,\middle|\, \lambda = (\lambda_1, \ldots, \lambda_{n+1}) \geq 0, \quad \sum_{i=1}^{n+1} \lambda_i = 1 \right\}.$$

It is clear that Y is a compact. Define the mapping $f : Y \to \mathbb{R}^n$ by the formula $f(y) = \sum_{i=1}^{n+1} \lambda_i x^i$, $y = (\lambda, x^1, \ldots, x^{n+1}) \in Y$. This mapping is continuous on Y. It follows from the Carathéodory theorem that $\operatorname{conv} X$ is an image of Y with the mapping f, that is, $\operatorname{conv} X = f(Y)$. But, as it is known, the image of a compact set is a compact set. □

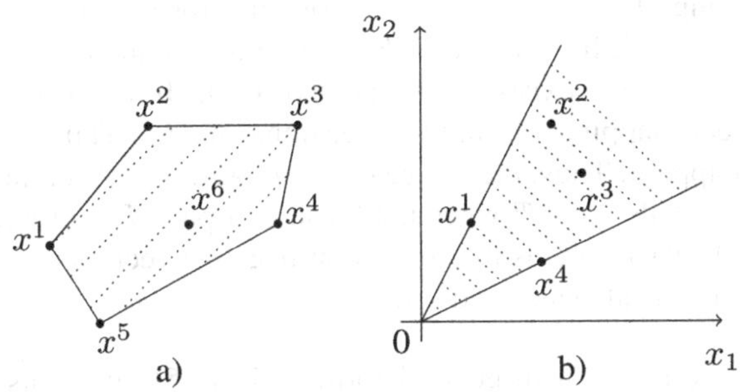

Figure 2.6. *a) Convex polyhedron. b) Polyhedral cone*

DEFINITION 2.9.– *The convex (conical) hull of a set consisting of a finite number of points is called the convex polyhedron (polyhedral cone) generated by these points.* Taking into account theorem 2.7, we can conclude that the convex polyhedron $X = \operatorname{conv}\{x^1, \ldots, x^m\}$ is of the form

$$X = \left\{ x \in \mathbb{R}^n \,\middle|\, x = \sum_{i=1}^{m} \lambda_i x^i,\ \lambda_i \geq 0;\ \sum_{i=1}^{m} \lambda_i = 1 \right\}, \qquad [2.4]$$

and the polyhedral cone $X = \text{cone}\left\{x^1, \ldots, x^m\right\}$ is of the form

$$X = \left\{ x \in \mathbb{R}^n \,\middle|\, x = \sum_{i=1}^{m} \lambda_i x^i,\ \lambda_i \geq 0,\ i = 1, \ldots, m \right\}. \qquad [2.5]$$

A set consisting of a finite number of points is compact. Therefore, we have the following corollary from the Carathéodory theorem.

THEOREM 2.9.– A convex polyhedron is compact.

The following statement is intuitively understandable, but its proof requires some effort.

THEOREM 2.10.– A polyhedral cone is closed.

PROOF.– Let X be of the form [2.5]. We first assume that the points $x^1, \ldots, x^m$ are linearly independent. Consider the linear subspace generated by these points

$$L = \left\{ x \in \mathbb{R}^n \,\middle|\, x = \sum_{i=1}^{m} \lambda_i x^i,\ \lambda_i \in R,\ i = 1, \ldots, m \right\}$$

and linear mapping $F : \mathbb{R}^m \to L$ of the form $F(\lambda) = \sum_{i=1}^{m} \lambda_i x^i$, $\lambda = (\lambda_1, \ldots, \lambda_m) \in \mathbb{R}^m$. It is clear that F is a one-to-one mapping. Then the inverse mapping $F^{-1} : L \to \mathbb{R}^m$ exists, and it is also linear. In addition, F and F^{-1} are continuous as linear mappings in finite-dimensional spaces. Thus, F is a one-by-one and continuous mapping. Thus, the image of any closed in $\mathbb{R}^m$ set under the mapping F is closed in L, and also in $\mathbb{R}^m$ since the linear subspace L is closed in $\mathbb{R}^m$. Notice that $X = F\left(\mathbb{R}^m_+\right)$, where $\mathbb{R}^m_+$ is the set of non-negative vectors in $\mathbb{R}^m$. Since the set $\mathbb{R}^m_+$ is closed, then X is also a closed set.

Let us prove the theorem in general form. If in [2.5] all points $x^1, \ldots, x^m$ are zero, then $X = \{0\}$ and the statement is trivial. Otherwise, any point in X can be represented as a non-negative combination of a linearly independent subsystem of points $x^1, \ldots, x^m$. This means that X is a combination of polyhedral cones, which are generated by all possible linearly independent subsystems of these points. The number of such cones is finite. They are closed. Thus, X is a closed set as a union of a finite number of closed cones. □

The intersection of any number of convex sets is a convex set. The same property has closed sets. Therefore, it is reasonable to introduce the following definition.

DEFINITION 2.10.– *The intersection of all closed convex sets that contain a set X is called the closed convex hull of X and is denoted by $\overline{\text{conv}}\, X$.*

THEOREM 2.11.– The following equality holds true:

$$\overline{\mathrm{conv}}\, X = \overline{\mathrm{conv}\, X}.$$

PROOF.– It is clear that $\overline{\mathrm{conv}}\, X \supseteq \mathrm{conv}\, X$ because in the process of formation $\mathrm{conv}\, X$ all convex sets are involved, not only closed. It follows that $\overline{\mathrm{conv}}\, X \supseteq \overline{\mathrm{conv}\, X}$. Conversely, $\overline{\mathrm{conv}\, X}$ is a convex closed set. That is why $\overline{\mathrm{conv}}\, X \subseteq \overline{\mathrm{conv}\, X}$, which completes the proof. □

THEOREM 2.12.– The convex hull of a compactum is compact.

PROOF.– Recall that in $\mathbb{R}^n$ any compact set is a bounded closed set. If $x \in \mathrm{conv}\, X$, where X is a compactum, then by theorem 2.8 (Carathéodory's theorem)

$$x = \sum_{i=1}^{n+1} \lambda_i x^i, \quad x^i \in X, \quad \lambda_i \geq 0, \quad \sum_{i=1}^{n+1} \lambda_i = 1.$$

That is why

$$\|x\| \leq \sum_{i=1}^{n+1} \lambda_i \left\|x^i\right\| \leq c,$$

where c is a constant such that $\|x\| \leq c$ for any $x \in X$. Hence, $\mathrm{conv}\, X$ is a bounded set. Let us show that it is closed. Let

$$x^k = \sum_{i=1}^{n+1} \lambda_{ik} x^{ik}, \quad x^{ik} \in X, \quad \sum_{i=1}^{n+1} \lambda_{ik} = 1.$$

Since the sequences λ_{ik} and x^{ik} are bounded, we can choose from them convergent subsequences. We can assume that $\lambda_{ik} \to \lambda_{i0}$ and $x^{ik} \to x^{i0} \in X$, since X is a compactum. Passing to the limit, we obtain

$$x^0 = \sum_{i=1}^{n+1} \lambda_{i0} x^{i0}, \quad x^{i0} \in X, \quad \sum_{i=1}^{n+1} \lambda_{i0} = 1.$$

This means that $x^0 \in \mathrm{conv}\, X$. The closeness of $\mathrm{conv}\, X$ is proved. □

2.3. Topological properties of convex sets

DEFINITION 2.11.– *A point $x \in X$ is called the interior point of the set X if there exists an $\varepsilon > 0$ such that $x + \varepsilon B \subseteq X$, where B is a unit ball in $\mathbb{R}^n$ with center at the origin, i.e. $B = \{x \in \mathbb{R}^n : \|x\| < 1\}$, $\|x\| = \langle x, x\rangle^{1/2}$. The set of such points is called the set of interior points of the set X and is denoted by $\mathrm{int}\, X$.*

DEFINITION 2.12.– *A point x is called the limit point of the set X if there exists a sequence of points $x^k \in X$ that converges to x. The set of all limit points of the set X is called its closure and is denoted by $\overline{X}$.*

LEMMA 2.8.– The closure $\overline{X}$ and the set of interior points of a convex set X are convex and, moreover, their affine hulls coincide $\operatorname{aff} X = \operatorname{aff} \overline{X}$.

PROOF.– If X is convex, then the fact that $x^1 \in \operatorname{int} X$, $x^2 \in \operatorname{int} X$ implies inclusions $x^1 + \varepsilon_1 B \subseteq X$, $x^2 + \varepsilon_2 B \subseteq X$. Let $\lambda_1 x^1 + \lambda_2 x^2$ be a convex combination of points x^1 and x^2. Then

$$\lambda_1(x^1 + \varepsilon_1 B) + \lambda_2(x^2 + \varepsilon_2 B) = \lambda_1 x^1 + \lambda_2 x^2 + (\lambda_1\varepsilon_1 + \lambda_2\varepsilon_2)B \subseteq X,$$

that is

$$\lambda_1 x^1 + \lambda_2 x^2 \in \operatorname{int} X.$$

If $x^1, x^2 \in \overline{X}$, then by definition there are sequences of points $x^{1k}, x^{2k} \in X$ such that $x^{1k} \to x^1$, $x^{2k} \to x^2$ as $k \to \infty$. Let $\lambda_1 x^{1k} + \lambda_2 x^{2k}$ be a convex combination of points x^{1k}, x^{2k}. Then

$$\lambda_1 x^1 + \lambda_2 x^2 = \lim_{k\to\infty} (\lambda_1 x^{1k} + \lambda_2 x^{2k}) \in \overline{X},$$

since $\lambda_1 x^{1k} + \lambda_2 x^{2k} \in X$ by virtue of convexity of X.

The affine set $\operatorname{aff} X$ is closed. Therefore, from $X \subset \operatorname{aff} X$ it follows that $\overline{X} \subset \operatorname{aff} X$. Hence $\operatorname{aff} \overline{X} \subset \operatorname{aff} X$. The reverse inclusion is obvious. □

Convex sets have such a property that in a certain sense they can always be placed in a subspace in which they already have interior points.

THEOREM 2.13.– A convex set X is in the space $\mathbb{R}^n$, or has interior points, or it is in a subspace of smaller dimension shifted by a vector.

PROOF.– Let $x^0 \in X$. Consider all vectors of the form $x - x^0, x \in X$. Among such vectors, there are $r \le n$ linearly independent: $x^1 - x^0, \ldots, x^r - x^0$. There are two possible cases:

a) $r = n$: In this case, there are n linearly independent vectors $x^i - x^0, x^i \in X; i = 1, \ldots, n$. Consider the set

$$S^n = \left\{x | x = \lambda_0 x^0 + \cdots + \lambda_n x^n \colon \lambda_i \ge 0, \lambda_0 + \cdots + \lambda_n = 1\right\}.$$

The set S^n is the called *n-dimensional simplex* generated by points $x^0, x^1, \ldots, x^n$. From theorem 2.6, we have $S^n \subseteq X$. If we prove that S^n has interior points, then X also has interior points. Let us prove that every point $x \in S^n$ with strictly positive coefficients λ_i belongs to int S^n. Consider the system of equations for $\lambda_i > 0, i = 1, \ldots, n$:

$$x - x^0 = \sum_{i=1}^{n} \lambda_i (x^i - x^0).$$

Since the vectors $x^i - x^0$ are linearly independent, this system has a unique solution $\lambda_i(x), i = 1, \ldots, n$, which depends continuously on x (we recall Cramer's formula for systems of equations with a non-degenerate determinant). Therefore, assuming x is equal to $\overline{x} = \bar{\lambda}_0 x^0 + \cdots + \bar{\lambda}_n x^n$, $\bar{\lambda}_i > 0$, we get $\lambda_i(\overline{x}) = \bar{\lambda}_i > 0, i = 1, \ldots, n$, and $\lambda_0 = 1 - \sum_{i=1}^{n} \bar{\lambda}_i > 0$. This implies that $\lambda_i(x) > 0, i = 1, \ldots, n$, for all x from some neighborhood of $\overline{x}$ and

$$\lambda_0(x) = 1 - \sum_{i=1}^{n} \lambda_i(x) > 0.$$

Therefore, for all points x from some neighborhood of $\bar{x}$ the inclusion

$$x = \sum_{i=1}^{n} \lambda_i(x) x^i \in S^n,$$

holds true. This proves the first part of the theorem.

b) $r < n$: Consider the subspace X^0, which consists of the vectors

$$y = \sum_{i=1}^{r} \alpha_i (x^i - x^0).$$

From the construction of the set X, we have $X - x^0 \subseteq X^0$, that is $X \subseteq x^0 + X^0$. □

The construction in the theorem subspace X^0 is r-dimensional and the set $X - x^0$ contains interior points in this subspace. This can be shown in the same way as in the proof of theorem 2.13. Further, X^0 does not depend on the choice of the point x^0 and vectors $x^i - x^0, i = 1, \ldots, r$. Really, any subspace containing $X - x^0$ must contain vectors $x^i - x^0$, and hence all X^0. It follows from this that X^0 is the intersection of all subspaces containing $X - x^0$. If the subspace X^1 contains $X - x^0$ for some $x^0 \in X$, then it contains $X - \overline{x}^0$ for any other point. Really, $x - \overline{x}^0 = (x - x^0) - (\overline{x}^0 - x^0)$, and since X^1 is a subspace, then it contains the difference of any two of its vectors. Thus, X^1 contains $X - x^0$ and $X - \overline{x}^0$ at the same time, i.e. X does not depend on the choice of x^0.

Now we can give a definition.

DEFINITION 2.13.– *A point x is called the relative interior point of a convex set X if $x + \text{Lin}\, X \cap (\varepsilon B) \subseteq X$, that is, x is contained in X together with a sphere of radius $\varepsilon > 0$, which lies in Lin X. The set of such points is called the relative interior of the convex set X and is denoted by ri X. The set X is called relatively open if $X = \text{ri}\, X$.*

LEMMA 2.9.– $\text{ri}\, \overline{X} = \text{ri}\, X$.

PROOF.– Since $\text{Lin}\, X$ is a closed set containing X, then $\text{Lin}\, X \supseteq \overline{X}$. We can see that $\text{Lin}\, X = \text{Lin}\overline{X}$. Obviously that $\text{ri}\, \overline{X} \supseteq \text{ri}\, X$. Prove the opposite inclusion. Let $x \in \text{ri}\, \overline{X}$, and let $e^i, \ldots, e^r$ be a basis in $\text{Lin}\, X$. Then for small ε, we have

$$y^k = x + \varepsilon \left(e^k - \frac{1}{r+1} e \right) \in \overline{X},\ k = 1, \ldots, r;$$

$$y^0 = x - \frac{\varepsilon}{r+1} e \in \overline{X},$$

where $e = e^1 + \cdots + e^r$. The vectors $y^k - y^0 = \varepsilon e^k$ are linearly independent and

$$x = \frac{1}{r+1} y^0 + \cdots + \frac{1}{r+1} y^r.$$

The last equality means that x is an interior point of a simplex, generated by points $y^0, \ldots, y^r$. If we take sufficiently close to y^k points $\overline{y}^k \in X$, then it turns out that $x^k \in X$ is the interior point of the simplex generated by the points $\overline{y}^k \in X$, and it is a relative interior point of X. Therefore, $\text{ri}\, \overline{X} \subseteq \text{ri}\, X$. □

THEOREM 2.14.– Let X be a convex set. If $x^1 \in \overline{X}, x^2 \in \text{ri}\, X$, then for all $\lambda \in (0, 1]$, the point $(1 - \lambda)x^1 + \lambda x^2 \in \text{ri}\, X$. Moreover, $\overline{X} = \overline{\text{ri}\, X}$.

PROOF.– If $x^2 \in \text{ri}\, X$, then $x^2 + \text{Lin}\, X \cap (\varepsilon B) \subseteq X$. From the convexity of X, it follows that

$$(1 - \lambda)x^1 + \lambda(x^2 + \text{Lin}\, X \cap (\varepsilon B)) =$$

$$(1 - \lambda)x^1 + \lambda x^2 + \text{Lin}\, X \cap (\lambda \varepsilon B) \subseteq \overline{X}.$$

That is $(1 - \lambda)x^1 + \lambda x^2 \in \text{ri}\, X$ since $\text{ri}\, X = \text{ri}\, \overline{X}$ by lemma 2.9.

Let $x^0 \in \overline{X}$, $x^k \in X$, $x^k \to x^0$, $\lambda_k \to 0$. Then $(1 - \lambda_k)x^k + \lambda_k y \in \text{ri}\, X$ if $y \in \text{ri}\, X$. That is why the point x^0 is a boundary point for $\text{ri}\, X$. Hence $\overline{X} \subseteq \overline{\text{ri}\, X}$. The reverse inclusion is obvious. □

DEFINITION 2.14.– *The dimension of the space Lin X is called the dimension of the convex set X. It is denoted by dim X.*

THEOREM 2.15.– If for convex sets X_1 and X_2 the condition

$$\mathrm{ri}\, X_1 \cap \mathrm{ri}\, X_2 \neq \emptyset$$

holds true, then

$$\mathrm{Lin}\, X_1 \cap \mathrm{Lin}\, X_2 = \mathrm{Lin}\,(X_1 \cap X_2),$$

$$\mathrm{ri}\, X_1 \cap \mathrm{ri}\, X_2 = \mathrm{ri}\,(X_1 \cap X_2).$$

PROOF.– Let us assume that $0 \in \mathrm{ri}\, X_1 \cap \mathrm{ri}\, X_2$. In this case, $\mathrm{Lin}\, X_1 \supseteq X_1$, $\mathrm{Lin}\, X_2 \supseteq X_2$. Therefore

$$\mathrm{Lin}\, X_1 \cap \mathrm{Lin}\, X_2 \supseteq X_1 \cap X_2,$$

and

$$\mathrm{Lin}\, X_1 \cap \mathrm{Lin}\, X_2 \supseteq \mathrm{Lin}\,(X_1 \cap X_2).$$

On the contrary, let $z \in \mathrm{Lin}\, X_1 \cap \mathrm{Lin}\, X_2$. Then for sufficiently small $\lambda > 0$, we have $\lambda z \in X_1$ and $\lambda z \in X_2$, since $0 \in \mathrm{ri}\, X_1 \cap \mathrm{ri}\, X_2$. It follows from here that $\lambda z \in X_1 \cap X_2$. Hence $\lambda z \in \mathrm{Lin}\,(X_1 \cap X_2)$. Since $\mathrm{Lin}\,(X_1 \cap X_2)$ is a subspace, then $z \in \mathrm{Lin}\,(X_1 \cap X_2)$.

Let us prove the second part of the statement. If $x \in \mathrm{ri}\, X_1 \cap \mathrm{ri}\, X_2$, then

$$x + \mathrm{Lin}\, X_1 \cap (\varepsilon B) \subseteq X_1, \quad x + \mathrm{Lin}\, X_2 \cap (\varepsilon B) \subseteq X_2,$$

so

$$(x + \mathrm{Lin}\, X_1 \cap (\varepsilon B)) \cap (x + \mathrm{Lin}\, X_2 \cap (\varepsilon B))$$

$$= x + \mathrm{Lin}\,(X_1 \cap X_2) \cap (\varepsilon B) \subseteq X_1 \cap X_2$$

and $x \in \mathrm{ri}\,(X_1 \cap X_2)$. That is why

$$\mathrm{ri}\,(X_1 \cap X_2) \supseteq \mathrm{ri}\, X_1 \cap \mathrm{ri}\, X_2.$$

Now let $x \in \mathrm{ri}\,(X_1 \cap X_2)$. Since $0 \in \mathrm{ri}\, X_j$, then $(1-\lambda)x \in \mathrm{ri}\, X_j, j = 1,2$ for $0 < \lambda \leq 1$. That is why

$$(1-\lambda)x \in \mathrm{ri}\, X_1 \cap \mathrm{ri}\, X_2.$$

Directing λ to zero, we get

$$x \in \overline{\mathrm{ri}\, X_1 \cap \mathrm{ri}\, X_2}.$$

Therefore

$$\mathrm{ri}\, X_1 \cap \mathrm{ri}\, X_2 \subseteq \mathrm{ri}\,(X_1 \cap X_2) \subseteq \overline{\mathrm{ri}\, X_1 \cap \mathrm{ri}\, X_2}.$$

Let $e^1, e^2, \ldots, e^r$ be a basis in $\text{Lin}\,(X_1 \cap X_2)$, $e = e^1 + \cdots + e^r$, $x \in \text{ri}\,(X_1 \cap X_2)$. Then, for a sufficiently small $\varepsilon > 0$, all points

$$y^k = x + \varepsilon\left(e^k - \frac{1}{r+1}e\right),\ k = 1, \ldots, r,\ y^0 = x - \frac{\varepsilon}{r+1}e$$

belong to $\text{ri}\,(X_1 \cap X_2)$. So they belong to $\overline{\text{ri}\,X_1 \cap \text{ri}\,X_2}$. At the same time

$$x = \frac{1}{r+1}y^0 + \cdots + \frac{1}{r+1}y^r$$

is an interior point of the simplex generated by the points $y^0, \ldots, y^k$, with respect to subspace $\text{Lin}\,(X_1 \cap X_2)$. If we take points $\overline{y}^k \in \text{ri}\,X_1 \cap \text{ri}\,X_2$ close enough to y^k, then x will be the interior point of the simplex generated by the points $\overline{y}^k$, and so it will belong to $\text{ri}\,X_1 \cap \text{ri}\,X_2$. It is thus proved that any point x from $\text{ri}\,(X_1 \cap X_2)$ belongs to $\text{ri}\,X_1 \cap \text{ri}\,X_2$. Together with the previous considerations this means that

$$\text{ri}\,(X_1 \cap X_2) = \text{ri}\,X_1 \cap \text{ri}\,X_2,$$

which was to be proved. □

THEOREM 2.16.– Let X be a convex set and let $x^0 \in \overline{X}$, but $x^0 \notin X$. Then in any neighborhood of x^0, there are points that do not belong to $\overline{X}$.

PROOF.– Take a point $y \in \text{ri}\,X$. Then the points of the ray $y + \lambda(x^0 - y)$, $\lambda \geq 0$, with $\lambda > 1$ do not belong to $\overline{X}$. In fact, if for $\lambda > 1$ the point $x^1 = y + \lambda(x^0 - y) \in \overline{X}$, then

$$x^0 = \frac{1}{\lambda}x^1 + \left(1 - \frac{1}{\lambda}\right)y \in \text{ri}\,X$$

by theorem 2.15, which contradicts the fact that $x^0 \notin X$. □

THEOREM 2.17.– Let X be an unbounded closed convex set in $\mathbb{R}^n$. Then:

1) for any point $x^0 \in X$, there exists a non-zero vector $h \in \mathbb{R}^n$ such that the ray

$$l^+_{x^0h} = \{x \in \mathbb{R}^n : x = x^0 + \alpha h,\ \alpha \geq 0\}$$

belongs to X;

2) if $l^+_{x^0h} \subset X$ for some $x^0 \in X$, then $l^+_{xh} \subset X$ for all $x \in X$; in other words, if a certain ray lies in X, then the ray with the origin at any point $x \in X$ in the same direction h also lies in X.

PROOF.– Let $x^0 \in X$. The set X is unbounded. Therefore, there is a sequence $x^k \in X$, $k = 1, 2\ldots$ such that $\|x^k\| \to \infty$. Take for $\alpha \geq 0$, $k = 1, 2, \ldots$

$$h^k = \frac{x^k - x^0}{\|x^k - x^0\|},\ \lambda_k = \frac{\alpha}{\|x^k - x^0\|},\ \overline{x}^k = \lambda_k x^k + (1 - \lambda_k)x^0.$$

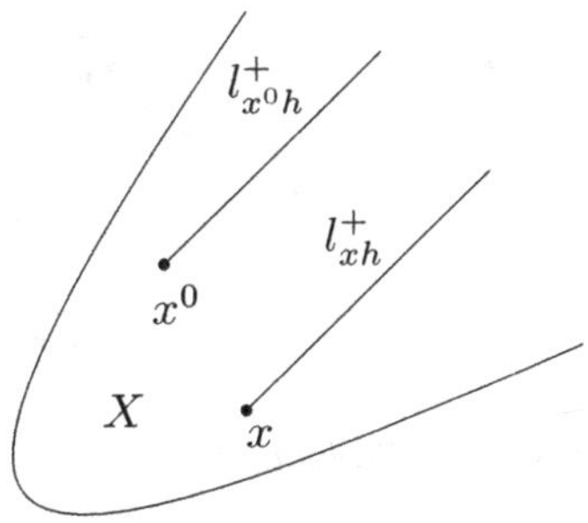

Figure 2.7. *Unbounded closed convex set*

Then $\|h^k\| = 1$ and we can assume that the sequence $\{h^k\}_{k=1}^{\infty}$ converges to a vector $h \neq 0$. For sufficiently large k, we have $0 \leq \lambda_k \leq 1$. Since the set X is convex, for such k the inclusion $\overline{x}^k \in X$ holds true. At the same time

$$\overline{x}^k = x^0 + \lambda_k(x^k - x^0) = x^0 + \alpha h^k.$$

Hence $\overline{x}^k \to x^0 + \alpha h$. Since the set X is closed we have $x^0 + \alpha h \in X$ for all $\alpha \geq 0$, that is $l^+_{x^0h} \subset X$.

Let $l^+_{x^0h} \subset X$ and $x \in X$. For all $\alpha \geq 0$ and $k = 1, 2, \ldots$ take

$$x^k = x^0 + (\alpha k)h, \; \overline{x}^k = \frac{1}{k}x^k + \left(1 - \frac{1}{k}\right)x.$$

Then $x^k \in l^+_{x^0h} \subset X$ and $\overline{x}^k \in X$ for all $k = 1, 2 \ldots.$. At the same time

$$\overline{x}^k = x + (x^k - x)/k = x + \alpha h + (x^0 - x)/k.$$

That is why $\overline{x}^k \to x + \alpha h$. Therefore, $x + \alpha h \in X$ for all $\alpha \geq 0$, that is $l^+_{xh} \subset X$. □

2.4. Theorems on separation planes and their applications

2.4.1. *Projection of a point onto a set*

DEFINITION 2.15.– *Projection of a point $a \in \mathbb{R}^n$ onto a set $X \subset \mathbb{R}^n$ is called a point $\pi_X(a) \in X$ such that $\|\pi_X(a) - a\| \leq \|x - a\|$ for all $x \in X$, that is, the closest point to the point a among all points x from X.*

If $a \in X$, then $\pi_X(a) = a$. If $a \notin X$ and the set X is open, then projection $\pi_X(a)$ does not always exist.

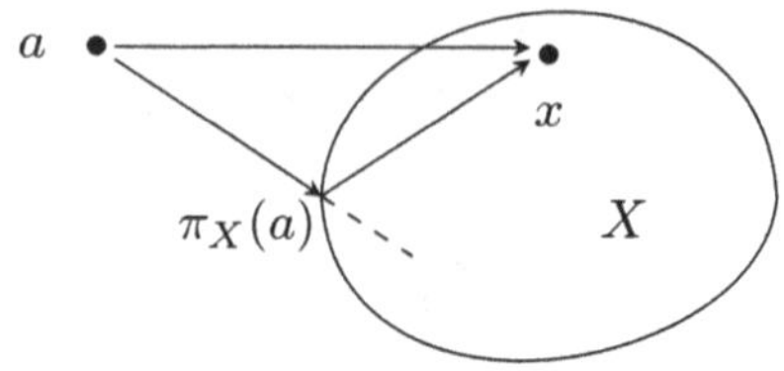

Figure 2.8. *Projection of a point onto a set*

LEMMA 2.10.– Let X be a closed convex set in $\mathbb{R}^n$ and let a point $a \notin X$. Then a projection $\pi_X(a)$ of the point $a \in \mathbb{R}^n$ onto the set X exists and has the following properties:

$$\langle \pi_X(a) - a, x - \pi_X(a) \rangle \geq 0 \quad \text{for all} \quad x \in X, \qquad [2.6]$$

$$\langle \pi_X(a) - a, x - a \rangle \geq \|\pi_X(a) - a\|^2 > 0 \quad \text{for all} \quad x \in X. \qquad [2.7]$$

Geometrically this means that the vectors $\pi_X(a) - a$ and $x - \pi_X(a)$ form a non-obtuse angle, and the angle between $\pi_X(a) - a$ and $x - a$ is an acute angle.

PROOF.– Take an arbitrary point $\hat{x} \in X$, take the number $R = \|\hat{x} - a\|$ and form the set

$$\hat{X} = \{x \in X \colon \|x - a\| \leq R\}.$$

This set is non-empty, closed and bounded. The continuous function $f(x) = \|x - a\|$ attains its minimum value on the set $\hat{X}$ at point x^*. This point is a point of minimum of the function $f(x) = \|x - a\|$ on the set X. Consequently, there exists a projection $x^* = \pi_X(a)$ of the point a onto the set X. For all $x \in X$ and $\lambda \in [0, 1]$ from convexity of the set X, we have

$$\begin{aligned} &\|x^* - a\|^2 \\ &\leq \|\lambda x + (1 - \lambda)x^* - a\|^2 = \|(x^* - a) + \lambda(x - x^*)\|^2 \\ &= \|x^* - a\|^2 + 2\lambda\langle x^* - a, x - x^* \rangle + \lambda^2\|x - x^*\|^2. \end{aligned}$$

Therefore

$$2\langle x^* - a, x - x^* \rangle + \lambda\|x - x^*\|^2 \geq 0.$$

We pass to the limit as $\lambda \to 0$ and obtain the first inequality. The second inequality is obtained if we add $\pm a$ to the second factor of the scalar product from the first inequality and take into account that $\pi_X(a) \neq a$ since $a \notin X$, $\pi_X(a) \in X$. $\square$

2.4.2. *Separation of two sets*

DEFINITION 2.16.– *The sets X_1 and X_2 from the space $\mathbb{R}^n$ are*

1) separated, if there exists a vector $p \in \mathbb{R}^n, p \neq 0$, and a number $\beta \in \mathbb{R}$ such that

$$\langle p, x^1\rangle \geq \beta \geq \langle p, x^2\rangle \quad \forall x^1 \in X_1, \forall x^2 \in X_2; \qquad [2.8]$$

2) properly separated, if there exist $p \in \mathbb{R}^n, p \neq 0$, and $\beta \in \mathbb{R}$ such that [2.8] holds true and, in addition,

$$\langle p, \hat{x}^1\rangle > \langle p, \hat{x}^2\rangle \text{ for some } \hat{x}^1 \in X_1,\ \hat{x}^2 \in X_2; \qquad [2.9]$$

3) strictly separated, if there exist $p \in \mathbb{R}^n, p \neq 0$, and $\beta \in \mathbb{R}$ such that

$$\langle p, x^1\rangle > \langle p, x^2\rangle \text{ for all } x^1 \in X_1,\ x^2 \in X_2; \qquad [2.10]$$

4) strongly separated, if there exist $p \in \mathbb{R}^n, p \neq 0$, and $\beta \in \mathbb{R}$ such that

$$\inf_{x^1 \in X_1} \langle p, x^1\rangle > \beta > \sup_{x^2 \in X_2} \langle p, x^2\rangle. \qquad [2.11]$$

Geometrically, this means that the sets X_1 and X_2 are located in different half-spaces

$$H^+_{p\beta} = \{x \in \mathbb{R}^n \colon \langle p, x\rangle \geq \beta\},$$

$$H^-_{p\beta} = \{x \in \mathbb{R}^n \colon \langle p, x\rangle \leq \beta\},$$

which are generated by the hyperplane

$$H_{p\beta} = \{x \in \mathbb{R}^n \colon \langle p, x\rangle = \beta\},\ p \neq 0.$$

We say that the hyperplane $H_{p\beta}$ *separates* X_1 and X_2, and the hyperplane $H_{p\beta}$ is called the *separating* hyperplane. With proper separation, the degenerate case is excluded, when both sets lie in the separating hyperplane. Strong separation means that the sets are at a positive distance from the hyperplane separating them, and hence from each other.

Recall that the distance between sets X_1 and X_2 is the number

$$\rho(X_1, X_2) = \inf_{x^1 \in X_1, x^2 \in X_2} \|x^1 - x^2\|.$$

THEOREM 2.18.– The convex sets $X_1 \subset \mathbb{R}^n$ and $X_2 \subset \mathbb{R}^n$ be strongly separated if and only if the distance between them is positive, that is,

$$\rho(X_1, X_2) = \inf_{x^1 \in X_1, x^2 \in X_2} \left\|x^1 - x^2\right\| > 0.$$

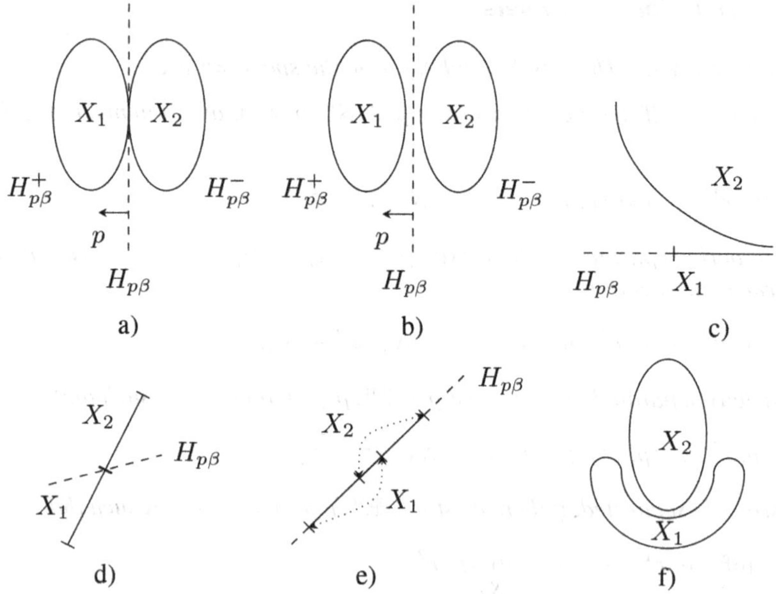

Figure 2.9. *Sets X_1 and X_2 are: a) properly separated; b) strongly separated; c) strictly separated; d) properly separated; e) separated; f) not separated*

PROOF.– Let the convex sets $X_1 \subset \mathbb{R}^n$ and $X_2 \subset \mathbb{R}^n$ be strongly separated. Then the following condition holds true:

$$\varepsilon = \inf_{x^1 \in X_1} \langle p, x^1 \rangle - \sup_{x^2 \in X_2} \langle p, x^2 \rangle = \inf_{x^1 \in X_1, x^2 \in X_2} \langle p, x^1 - x^2 \rangle > 0.$$

It follows from the Cauchy–Bunyakovsky inequality that

$$\varepsilon \leq \langle p, x^1 - x^2 \rangle \leq \|p\| \|x^1 - x^2\|$$

for all $x^1 \in X_1, x^2 \in X_2$. Hence

$$\rho(X_1, X_2) \geq \varepsilon / \|p\| > 0.$$

Analogously we get $\rho(X_1, H_{p\beta}) > 0$ and $\rho(X_2, H_{p\beta}) > 0$.

Let $\rho(X_1, X_2) > 0$. Consider the set $X = \overline{X_1 - X_2}$. It is convex and closed. From the condition $\rho(X_1, X_2) > 0$, it follows that $0 \notin X$. Let $p = \pi_X(0)$ be the projection of the point 0 on the set X. It follows from the inequality [2.7] that $\langle p, x\rangle \geq \|p\|^2$ for all $x \in X$. Hence

$$\inf_{x^1 \in X_1, x^2 \in X_2} \langle p, x^1 - x^2\rangle > 0.$$

Therefore

$$\inf_{x^1 \in X_1} \langle p, x^1\rangle > \sup_{x^2 \in X_2} \langle p, x^2\rangle.$$

If β lies between these numbers, then inequality [2.11] holds true. That is, the sets $X_1 \subset \mathbb{R}^n$ and $X_2 \subset \mathbb{R}^n$ are strongly separated. □

COROLLARY 2.3.– If the closed convex sets $X_1 \subset \mathbb{R}^n$ and $X_2 \subset \mathbb{R}^n$ do not intersect and at least one of them is bonded, then they are strongly separated.

PROOF.– It is easy to verify that under these conditions $\rho(X_1, X_2) > 0$. □

Note that the condition of boundedness here is significant and examples are given.

As a corollary, we have the following statement.

COROLLARY 2.4.– (*Minkowski's theorem on separation of a point and a set*) Let $X \subset \mathbb{R}^n$ be a closed convex set, and let a be a point in $\mathbb{R}^n$ that does not belong to X. Then there exist $p \in \mathbb{R}^n$ and $\beta \in \mathbb{R}$ such that

$$\inf_{x \in X} \langle p, x\rangle > \beta > \langle p, a\rangle.$$

In other words, it is asserted that there exists a hyperplane

$$H_{p\beta} = \{x \in \mathbb{R}^n\colon \langle p, x\rangle = \beta\}$$

such that the set X is in one half-space generated by $H_{p\beta}$, and the point a is inside another half-space:

$$X \subset H^+_{p\beta} = \{x \in \mathbb{R}^n\colon \langle p, x\rangle \geq \beta\},$$

$$a \in \operatorname{int} H^-_{p\beta} = \{x \in \mathbb{R}^n\colon \langle p, x\rangle \leq \beta\}.$$

At the same time, the hyperplane

$$H_{p\beta_1} = \{x \in \mathbb{R}^n\colon \langle p, x\rangle = \langle p, a\rangle = \beta_1\},$$

passing through the point a defines a half-space $H^+_{p\beta_1}$ such that

$$X \subset \operatorname{int} H^+_{p\beta_1}.$$

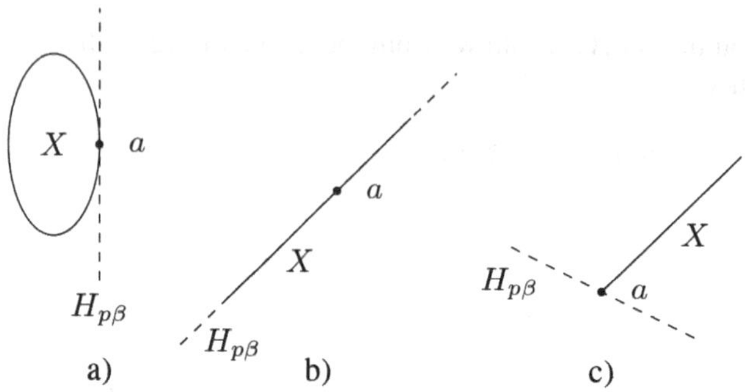

Figure 2.10. *a), c) Properly supporting hyperplanes; b) supporting hyperplane*

DEFINITION 2.17.– *A hyperplane $H_{p\beta}$ is called supporting of a set $X \subset \mathbb{R}^n$ at a point $a \in \partial X = \overline{X} \setminus \operatorname{int} X$ if X is contained in one of the half-spaces generated by this hyperplane, and it contains a point a, that is,*

$$\langle p, x\rangle \geq \beta = \langle p, a\rangle \text{ for all } x \in X. \quad [2.12]$$

DEFINITION 2.18.– *A hyperplane $H_{p\beta}$ is called properly supporting of a set $X \subset \mathbb{R}^n$ at a point $a \in \partial X = \overline{X} \setminus \operatorname{int} X$ if it supports X at the point a but does not completely contain X, i.e.*

$$\langle p, \hat{x}\rangle > \beta \text{ for some } \hat{x} \in X. \quad [2.13]$$

The half-space generated by a supporting (proper supporting) hyperplane to X at a point a and containing X is also called supporting (proper supporting) to X at the point a.

THEOREM 2.19.– 1) At any boundary point $a \in \partial X = \overline{X} \setminus \operatorname{int} X$ of a convex set $X \subset \mathbb{R}^n$, there exists a supporting hyperplane.

2) At any relative boundary point $a \in r\partial X = \overline{X} \setminus \operatorname{ri} X$ of a convex set $X \subset \mathbb{R}^n$, there exists a proper supporting hyperplane.

PROOF.– Prove the second statement, provided that the set X is closed. Condition $a \in r\partial X = \overline{X} \setminus \operatorname{ri} X$ means that there is a sequence of points $a^k \in \operatorname{aff} X \setminus X, k = 1, 2, \ldots$, which converges to a. Define the vectors

$$p^k = \frac{\pi_X(a^k) - a^k}{\|\pi_X(a^k) - a^k\|}, \; k = 1, 2, \ldots$$

It follows from [2.7] that

$$\langle p^k, x\rangle > \langle p^k, a^k\rangle \text{ for all } x \in X. \quad [2.14]$$

Since $\|p^k\| = 1$, then we can assume that $p^k \to p \neq 0$. Then, passing to the limit, we get [2.12]. Let $\hat{x} \in \operatorname{ri} X$, that is, $U_\varepsilon(\hat{x}) \cap \operatorname{aff} X \subset X$ for some $\varepsilon > 0$. Consider the point $x = \hat{x} - \varepsilon p$. Since $p^k \in \operatorname{Lin} X$, and the linear space $\operatorname{Lin} X$ is closed, then $p \in \operatorname{Lin} X$. Hence $x \in \operatorname{aff} X$. In addition, $x \in U_\varepsilon(\hat{x})$ since $\|p\| = 1$. Hence $x \in U_\varepsilon(\hat{x}) \cap \operatorname{aff} X \subset X$. Substitute this point to [2.12]. We get $\langle p, x\rangle = \langle p, \hat{x}\rangle - \varepsilon \geq \beta$, that is $\langle p, \hat{x}\rangle > \beta$. That is why [2.13] holds true for arbitrary $\hat{x} \in \operatorname{ri} X$. Statement (2) for closed sets is proved.

In general cases, we need to apply similar arguments to closure of the set X taking into account that $\operatorname{ri} X = \operatorname{ri} \overline{X}$ and $r\partial X = r\partial \overline{X}$. To prove statement (1), we consider two cases: $\operatorname{int} X = \emptyset$, $\operatorname{int} X \neq \emptyset$. If $\operatorname{int} X = \emptyset$, then $\operatorname{aff} X \neq \mathbb{R}^n$. Then $\operatorname{aff} X$ is the intersection of some number of hyperplanes, each of which is supporting X at any point $a \in \partial X = \overline{X} \setminus \operatorname{int} X$. If $\operatorname{int} X \neq \emptyset$, then $\partial X = r\partial X$ and we can use statement (2). □

THEOREM 2.20.– If the convex sets $X_1 \subset \mathbb{R}^n$, $X_2 \subset \mathbb{R}^n$ do not intersect, then they are separated.

This statement is a simplified version of the following fundamental theorem of convex analysis.

THEOREM 2.21.– (*Fenchel's theorem on proper separation of sets*) Convex sets $X_1 \subset \mathbb{R}^n, X_2 \subset \mathbb{R}^n$ are proper separated if and only if their relative interiors do not intersect:

$$\operatorname{ri} X_1 \cap \operatorname{ri} X_2 = \emptyset.$$

PROOF.– Let $\operatorname{ri} X_1 \cap \operatorname{ri} X_2 = \emptyset$. Consider the set $X = \operatorname{ri} X_1 - \operatorname{ri} X_2$. This set is convex and $0 \notin X$. There are two possible cases: $0 \notin \overline{X}$, $0 \in \overline{X} \subset r\partial X$. Apply in the first case corollary 2.4, and in the second case apply assertion (2) of theorem 2.19. We find that there exists a vector p for which $\langle p, x\rangle \geq \langle p, 0\rangle$, $x \in X$, and for some $\hat{x} \in X$ strict inequality takes place. This means that

$$\langle p, x^1\rangle \geq \langle p, x^2\rangle \ \forall x^1 \in \operatorname{ri} X_1, \ \forall x^2 \in \operatorname{ri} X_2, \qquad [2.15]$$

and for some $\hat{x}^1 \in \operatorname{ri} X_1$, $\hat{x}^2 \in \operatorname{ri} X_2$ strict inequality holds true, that is, [2.4] holds true. Since transition to the boundary does not change non-strict inequalities, then the sets $\operatorname{ri} X_1, \operatorname{ri} X_2$ can be replaced by their closure. But $X_1 \subset \overline{\operatorname{ri} X_1}$, $X_2 \subset \overline{\operatorname{ri} X_2}$. Hence

$$\langle p, x^1\rangle \geq \langle p, x^2\rangle, \ \forall x^1 \in X_1, \ \forall x^2 \in X_2.$$

Therefore

$$\inf_{x^1 \in X_1} \langle p, x^1\rangle \geq \sup_{x^2 \in X_2} \langle p, x^2\rangle.$$

If the value β lies between these numbers (or equals one of them), condition [2.8] is satisfied. This proves the theorem in one direction.

Now let the sets X_1, X_2 be properly separated, that is, for some p and β the relations [2.8] and [2.9] hold true. Suppose that $x \in \operatorname{ri} X_1 \cap \operatorname{ri} X_2$. Take a small $\alpha < 0$ such that

$$x^1 = x + \alpha(\hat{x}^1 - x) \in X_1, \quad x^2 = x + \alpha(\hat{x}^2 - x) \in X_2,$$

where $\hat{x}^1, \hat{x}^2$ are taken from [2.9]. Then $\langle p, x^1 \rangle < \langle p, x^2 \rangle$, which contradicts [2.8]. Therefore, $\operatorname{ri} X_1 \cap \operatorname{ri} X_2 = \emptyset$. □

2.5. Systems of linear inequalities and equations

As an example of application of separation theorems, we present the following statement.

THEOREM 2.22.– (*Minkowski–Farkas theorem*) Let A be a matrix of dimension $m \times n$, and let $b \in \mathbb{R}^m$ be a vector. Only one of the systems has solutions:

$$Ax = b, \qquad x \geq 0, \qquad x \in \mathbb{R}^n; \tag{2.16}$$

$$pA \geq 0, \quad \langle p, b \rangle < 0, \quad p \in \mathbb{R}^m. \tag{2.17}$$

PROOF.– Assuming that systems [2.16] and [2.17] have solutions $x \in \mathbb{R}^n$ and $p \in \mathbb{R}^m$ simultaneously, we get a contradiction:

$$0 > \langle p, b \rangle = \langle p, Ax \rangle = \langle pA, x \rangle \geq 0. \tag{2.18}$$

Assume now that system [2.16] has no solutions. This means that the vector b does not belong to the set

$$Y = \{y \in \mathbb{R}^m \colon y = Ax, x \geq 0\}.$$

This set is closed and convex. By the Minkowski theorem, there exists a vector $p \in \mathbb{R}^m$ such that

$$\langle p, y \rangle > \langle p, b \rangle \quad \text{for all } y \in Y.$$

In other words

$$\langle p, Ax \rangle = \langle pA, x \rangle > \langle p, b \rangle \quad \text{for all } x \geq 0.$$

Hence $pA \geq 0$, since coordinates of the vector x can be arbitrarily large. If we take $x = 0$, we obtain $\langle p, b \rangle < 0$. So p is a solution of the system [2.17]. □

From the Minkowski–Farkas theorem, we can derive a series of similar results. Here is one of them.

THEOREM 2.23.– Let A be a matrix of $m \times n$ dimension, and let $b \in \mathbb{R}^m$ be a vector. Only one of the systems has solutions:

$$Ax \leq b, \quad x \in \mathbb{R}^n; \tag{2.19}$$

$$pA = 0, \quad \langle p, b \rangle < 0, \quad p \geq 0, \quad p \in \mathbb{R}^m. \tag{2.20}$$

PROOF.– Similarly to [2.18], we show that systems cannot have a solution at the same time. Assume now that the system [2.20] has no solutions. Then there is no solution to the system

$$pA = 0, \quad \langle p, b \rangle = -1, \quad p \geq 0.$$

This system can be represented in the form [2.16]

$$\begin{bmatrix} A^\top \\ b \end{bmatrix} p = \begin{bmatrix} 0 \\ -1 \end{bmatrix}, \quad p \geq 0.$$

By the preceding theorem, there exists a vector $(x, \lambda) \in \mathbb{R}^n \times \mathbb{R}$ such that

$$(x, \lambda) \begin{bmatrix} A^\top \\ b \end{bmatrix} \geq 0, \quad \langle (x, \lambda); (0, -1) \rangle < 0.$$

That is

$$Ax + \lambda b \geq 0, \quad \lambda > 0.$$

It follows that $\bar{x} = -x/\lambda$ is a solution of the system [2.19]. □

As a result, we obtain one important property of linear programming problems.

THEOREM 2.24.– If in the linear minimization (maximization) problem the admissible set is non-empty, and the objective function is bounded from below (from above) on the admissible set, then the problem has solutions.

PROOF.– Let a matrix A of dimension $m \times n$, a vector $c \in \mathbb{R}^n$ and a vector $b \in \mathbb{R}^m$ be given. Consider the linear programming problem

$$\langle c, x \rangle \to \min, \quad x \in X = \{x \in \mathbb{R}^n : Ax \geq b\}.$$

Let $X \neq \emptyset$ and $\alpha = \inf_{x \in X} \langle c, x \rangle > -\infty$. Assume that the problem has no solutions. That is, there is no solution of the system

$$Ax \geq b, \langle c, x \rangle \leq \alpha.$$

This can be written as follows:

$$\begin{bmatrix} -A \\ c \end{bmatrix} x \leq \begin{bmatrix} -b \\ \alpha \end{bmatrix}.$$

Then, according to the preceding theorem, there exists a vector $(p, \lambda) \in \mathbb{R}^m \times \mathbb{R}$ such that

$$(p, \lambda) \begin{bmatrix} -A \\ c \end{bmatrix} = 0, \quad \langle (p, \lambda); (-b, \alpha) \rangle < 0, \quad (p, \lambda) \geq 0.$$

That is

$$pA = \lambda c, \quad \langle p, b \rangle > \lambda \alpha, \quad p \geq 0, \lambda \geq 0.$$

From this for any $x \in X$, we get

$$\lambda \langle c, x \rangle = \langle pA, x \rangle = \langle p, Ax \rangle \geq \langle p, b \rangle > \lambda \alpha.$$

Hence $\lambda > 0$ and $\inf_{x \in X} \langle c, x \rangle \geq \langle p, b \rangle / \lambda > \alpha$, which contradicts the definition of α. □

REMARK 2.1.– We emphasize that in this theorem the admissible set is not necessarily bounded, while for nonlinear problems this case is not permitted. For example, the function $f(x) = e^x$ is bounded from below on $\mathbb{R}$, but does not attain the minimum.

2.6. Extreme points of a convex set

DEFINITION 2.19.– *A point x of a convex set $X \subset \mathbb{R}^n$ is called extreme if it cannot be presented as*

$$x = \lambda x^1 + (1 - \lambda) x^2, \text{ where } x^1, x^2 \in X, \ x^1 \neq x^2, \ 0 < \lambda < 1. \qquad [2.21]$$

The set of all extreme points of a set X is denoted by $E(X)$.

Thus, a point x is extreme in X if it cannot be placed in the middle of a segment whose ends lie in X. For example, in a triangle the extreme points are its vertices, in a ray this is the beginning, and in a circle the extreme points are all points of the circle.

We now provide a lemma which is a useful tool for proving the following basic theorems of the theory of extreme points. The formulation of the lemma is based on theorem 2.19.

LEMMA 2.11.– Let X be a closed convex set in $\mathbb{R}^n$, and let $H = H_{p\beta}$ be a properly supporting X at the point $\hat{x} \in r\partial X = X \setminus \operatorname{ri} X$ hyperplane, that is, the conditions are satisfied

$$\langle p, x \rangle \geq \beta = \langle p, \hat{x} \rangle \quad \text{for all } x \in X, \qquad [2.22]$$

$$\langle p, \bar{x} \rangle > \beta \quad \text{for some } \bar{x} \in X. \qquad [2.23]$$

Take $\hat{X} = X \cap H$. Then:

1) any extreme point in $\hat{X}$ is an extreme point in X, that is, $E(\hat{X}) \subset E(X)$;

2) $\dim \hat{X} < \dim X$.

PROOF.– 1) Let $x \in E(\hat{X})$, but $x \notin E(X)$, that is, x can be presented as [2.21]. Using [2.22], we get

$$\beta = \langle p, x \rangle = \lambda \langle p, x^1 \rangle + (1 - \lambda) \langle p, x^2 \rangle \geq \lambda \beta + (1 - \lambda) \beta = \beta.$$

Hence $\langle p, x^1 \rangle = \langle p, x^2 \rangle = \beta$. That is $x^1, x^2 \in \hat{X} = X \cap H$. Together with [2.21], this means that $x \notin E(\hat{X})$. This contradiction proves that $x \in E(X)$. That is $E(\hat{X}) \subset E(X)$.

2) Take $M = \operatorname{aff} X$, $\hat{M} = \operatorname{aff} \hat{X}$. Then $\hat{M} \subset M$, $\hat{M} \subset H$, since $\hat{X} \subset X$, $\hat{X} \subset H$. Suppose that $\hat{M} = M$. Then $X \subset M = \hat{M} \subset H$, that is $\beta = \langle p, x \rangle \ \forall x \in X$, which contradicts [2.23]. Hence $\hat{M} \neq M$. Parallel subspaces $L = \operatorname{Lin} X$ and $\hat{L} = \operatorname{Lin} \hat{X}$ are related by the same relations $\hat{L} \subset L$, $\hat{L} \neq L$. Therefore, the basis in L has at least one vector more than the basis in $\hat{L}$, i.e. $\dim \hat{L} < \dim L$. By definition, $\dim X = \dim L$, $\dim \hat{X} = \dim \hat{L}$. So $\dim \hat{X} < \dim X$. □

THEOREM 2.25.– (*Criterion of existence of extreme points*) Let X be a closed convex set in $\mathbb{R}^n$. Then X has at least one extreme point if and only if X does not include a straight line, that is, the sets of the form

$$l_{x^0 h} = \{x \in \mathbb{R}^n \colon x = x^0 + \alpha h, \ \alpha \in \mathbb{R}\},$$

where $x^0 \in \mathbb{R}^n$, $h \in \mathbb{R}^n$, $h \neq 0$.

PROOF.– 1) Let $x \in E(X)$, but $l_{x^0 h} \subset X$ for some x^0 and $h \neq 0$. Then by theorem 2.17 we have $l_{xh} \subset X$, so $x_1 = x + h \in X$ and $x_2 = x - h \in X$. Herewith, $x = 0,5x_1 + 0,5x_2$, $x_1 \neq x_2$, that is $x \notin E(X)$. This proves the statement of the theorem in one direction.

2) Suppose now that X does not contain lines. We show that $E(X) \neq \emptyset$ by the method of mathematical induction with respect to the dimension of the set X. If $\dim X = 0$, then $X = \{\hat{x}\}$ is one-point set $E(X) = \{\hat{x}\} \neq \emptyset$. Let the statement be true for $\dim X < m$ and $\dim X = m$. Take any point $\hat{x} \in r\partial X$. Let $H = H_{p\beta}$ be properly supporting the set X at point $\hat{x}$ hyperplane. Take $\hat{X} = X \cap H$. This set is closed, convex, and does not contain straight lines. Moreover $\dim \hat{X} < m$. By assumption of the theorem, $E(\hat{X}) \neq \emptyset$. But $E(\hat{X}) \subset E(X)$, that is why $E(X) \neq \emptyset$. □

THEOREM 2.26.– (*Minkowski's theorem on a convex compact*) Let X be a convex compact (closed bounded set) in $\mathbb{R}^n$. Then $X = \operatorname{conv} E(X)$, that is, X coincides with the convex hull of the set of its extreme points.

PROOF.– We prove the assertion by the method of mathematical induction on the dimension of the set X. If $\dim X = 0$, then the theorem is obvious. Let the statement be true for $\dim X < m$ and let $\dim X = m$.

Let $H = H_{p\beta}$ be properly supporting the set X at a point $\hat{x} \in r\partial X$ hyperplane. Take $\hat{X} = X \cap H$. In this case, the set $\hat{X}$ is a convex compact and $\dim \hat{X} < m$. Then by the assumption of induction $\hat{X} = \operatorname{conv} E(\hat{X})$. We have $\hat{x} \in \operatorname{conv} E(\hat{X})$. But $E(\hat{X}) \subset E(X)$, therefore $\hat{x} \in \operatorname{conv} E(X)$. Hence $r\partial X \subset \operatorname{conv} E(X)$.

Consider now an arbitrary point $\hat{x} \in \operatorname{ri} X$ and a vector $h \in \operatorname{Lin} X$. Then the line $l_{\hat{x}h}$ lies in $\operatorname{aff} X$. The intersection of this line with X forms a segment with ends on the relative boundary of X. That is $l_{\hat{x}h} \cap X = \operatorname{conv}\{x^1, x^2\}$, $x^1, x^2 \in r\partial X$. Hence $\hat{x} \in \operatorname{conv}\{x^1, x^2\} \subset \operatorname{conv}(r\partial X) \subset \operatorname{conv}(\operatorname{conv} E(X)) = \operatorname{conv} E(X)$. That is why $\operatorname{ri} X \subset \operatorname{conv} E(X)$, $X = r\partial X \cup \operatorname{ri} X \subset \operatorname{conv} E(X)$. The inverse inclusion of $\operatorname{conv} E(X) \subset X$ is obvious because $E(X) \subset X$ and X is a convex set. □

DEFINITION 2.20.– *A polyhedron is called the set of solutions of a system of finite number of linear inequalities, i.e. the intersection of a finite number of half-spaces:*

$$X = \left\{x \in \mathbb{R}^n \colon \langle a^i, x\rangle \leq b_i, i \in I = \{1, \dots, m\}\right\}, \qquad [2.24]$$

or

$$X = \{x \in \mathbb{R}^n \colon Ax \leq b\},$$

where $b = (b_1, \dots, b_m) \in \mathbb{R}^m$, *$A$ is a matrix of dimension $m \times n$ with rows* $a^1, a^2, \dots, a^m \in \mathbb{R}^n$.

THEOREM 2.27.– For a point $\hat{x}$ to be an extreme point of a polyhedron X given by system [2.24] of linear inequalities, it is necessary and sufficient that the set

$$I(\hat{x}) = \left\{i \colon \langle a^i, \hat{x}\rangle = b_i, i \in I\right\}$$

contains a subset I_0 of dimension n such that vectors $a^i, i \in I_0$ are linearly independent.

PROOF.– Necessity: let the set $\{a_i \colon i \in I(\hat{x})\}$ contain less than n linearly independent elements. Then, based on the well-known theorems of linear algebra, the system of linear equations

$$\langle a^i, x\rangle = 0, i \in I(\hat{x}),$$

has a non-zero solution $\overline{x}$. It follows from the definition of the set $I(\hat{x})$ that

$$\langle \hat{x} \pm \varepsilon\overline{x}, a^i \rangle = b_i, i \in I(\hat{x}),$$

$$\langle \hat{x} \pm \varepsilon\overline{x}, a^i \rangle < b_i, i \in I \backslash I(\hat{x}),$$

for a sufficiently small $\varepsilon > 0$, such that

$$\hat{x} + \varepsilon\overline{x} \in X, \quad \hat{x} - \varepsilon\overline{x} \in X,$$

$$\hat{x} = \frac{1}{2}(\hat{x} + \varepsilon\overline{x}) + \frac{1}{2}(\hat{x} - \varepsilon\overline{x}) \in X,$$

that is, $\hat{x}$ is not an extreme point of the set X.

Sufficiency: let the point $\hat{x} \in X$, the dimension of I_0 be equal to n and for $i \in I_0$ vectors let a^i be linearly independent. Then the system of inequalities describing the set X can be written in the following form:

$$\langle \hat{x}, a^i \rangle = b_i, \quad i \in I_0, \qquad [2.25]$$

$$\langle \hat{x}, a^i \rangle \leq b_i, \quad i \in I \backslash I_0. \qquad [2.26]$$

Suppose that

$$\hat{x} = 0,5x^1 + 0,5x^2, \quad x^1 \in X, \quad x^2 \in X, \quad x_1 \neq x_2. \qquad [2.27]$$

Since $x^1, x^2 \in X$, then by definition the following inequalities hold true:

$$\langle x^k, a^i \rangle \leq b_i, \quad i \in I_0, \; k = 1, 2. \qquad [2.28]$$

According to conditions [2.25] and [2.28], the relation [2.27] is satisfied only if

$$\langle x^k, a^i \rangle = b_i, \quad i \in I_0, \quad k = 1, 2. \qquad [2.29]$$

Consequently, two different points satisfy the system n of linearly independent equations [2.29]. This is impossible due to the well-known theorems. □

From the above two theorems, the following theorem follows.

THEOREM 2.28.– A bounded polyhedron, given by a finite system of linear inequalities [2.24], is a convex hull of its extreme points, the number of which is finite.

2.7. Exercises

1) Prove that a set X is convex if and only if

$$\lambda_1 X + \lambda_2 X = (\lambda_1 + \lambda_2)X$$

for all $\lambda_1 \geq 0, \lambda_2 \geq 0$.

2) Let X be a closed set, and let there exist, for any points $x^1 \in X$ and $x^2 \in X$, a number $\lambda \in (0,1)$ such that $\lambda x^1 + (1-\lambda)x^2 \in X$. Prove that the set X is convex. Give an example that shows that the condition of closedness of X is significant here.

3) Prove the convexity of the following sets in $\mathbb{R}^2$:

a) $X_1 = \{x \in \mathbb{R}^2 | x_2 \geq x_1^2\}$;

b) $X_2 = \{x \in \mathbb{R}^2 | x_1 x_2 \geq 1, x_1 > 0\}$;

c) $X_3 = \{x \in \mathbb{R}^2 | sin(x_1) \geq x_2, x_1 \in [0, \pi]\}$;

d) $X_4 = \{x \in \mathbb{R}^2 | x_2 \geq \exp(x_1)\}$:

e) $X_5 = \{x \in \mathbb{R}^2 | 1 \geq x_1^3 + x_2^3, x_1 + x_2 \geq k\}$.

4) Prove that a ball in $\mathbb{R}^n$ is a convex set.

5) Let A be a non-negative definite matrix of dimension $n \times n$. Prove that the set

$$X = \{x \in \mathbb{R}^n \, | \langle Ax, x \rangle \leq \alpha\}$$

is convex for arbitrary $\alpha \geq 0$, and X is a linear subspace, if $\alpha = 0$.

6) Prove that the set

$$X = \left\{x \in \mathbb{R}^3 \, | x_1^2 - 2x_1 x_3 + x_2^2 \leq 0, \; x_1 \geq 0\right\}$$

is a convex cone. Draw this cone.

7) Let $X_1, \ldots, X_m$ be convex sets in $\mathbb{R}^{n_1}, \ldots, \mathbb{R}^{n_m}$. Prove that the Cartesian product $X = X_1 \times \ldots \times X_m$ is also a convex set.

8) Let A: $\mathbb{R}^n \to \mathbb{R}^m$ be a linear mapping, and let $X \subset \mathbb{R}^n$ and $Y \subset \mathbb{R}^m$ be convex sets. Prove that the sets

$$A(X) = \{y \in \mathbb{R}^m \, | y = Ax \;\text{ for some } x \in X\},$$

$$A^{-1}(Y) = \{x \in \mathbb{R}^n \, | Ax \in Y\},$$

that is, the image of X and the prime of Y, are also convex.

9) Draw the sum of sets on a plane

$$X_1 = \{x \in \mathbb{R}^2 \, | \, |x_i| \leq 1, \; i = 1, 2\},$$

$$X_2 = \{x \in \mathbb{R}^2 \, | x_1^2 + x_2^2 \leq 1\}.$$

10) Prove that the sum of an open and arbitrary set is open.

11) Give an example of two closed convex sets such that their sum is not closed.

12) Give an example of two closed convex cones such that their sum is not closed.

13) The *projection* of the set $X \subset \mathbb{R}^n \times \mathbb{R}^m$ (on the space of the first n coordinates) is called the set

$$P(X) = \{y \in \mathbb{R}^n \mid (y, z) \in X \text{ for some } z \in \mathbb{R}^m\}.$$

Prove that the projection of a convex set is convex. Prove that the projection of a compactum is compact. Give an example of a closed convex set such that its projection is not closed.

14) Let X_1, X_2 be convex sets in $\mathbb{R}^n \times \mathbb{R}^m$. Prove that the set

$$X = \{(y, z) \in \mathbb{R}^n \times \mathbb{R}^m \,|(y, z^1) \in X_1, (y, z^2) \in X_2,$$

$$z = z^1 + z^2 \quad \text{for some } z^1, z^2 \in \mathbb{R}^m\},$$

which is called the *partial sum* of the sets X_1, X_2, is convex. Note that in the cases $n = 0$ and $m = 0$, this set will have the forms

$$X = X_1 + X_2, \quad X = X_1 \cap X_2.$$

15) Let X_1, X_2 be convex sets in $\mathbb{R}^n$. Prove that the set

$$X = \bigcup_{0 \le \lambda \le 1} (\lambda X_1 \cap (1 - \lambda) X_2),$$

which is called the *inverse sum* of sets X_1, X_2, is convex.

16) Let X_1, X_2 be convex sets in $\mathbb{R}^n$. Prove that the sets

$$X = \bigcap_{x \in X_2} \bigcup_{\lambda \ge 1} (\lambda X_1 + (1 - \lambda) x),$$

$$Y = \bigcup_{x \in X_2} \bigcup_{\lambda \ge 1} (\lambda X_1 + (1 - \lambda) x),$$

which are called the *shadow* and *semi-shadow* of the set X_1 with respect to the set ("source of light") X_2, are convex.

17) Let $X_1, \dots, X_m$ be convex sets in $\mathbb{R}^n$, and let Λ be a convex set in $\mathbb{R}^m$, such that $\Lambda \subset \mathbb{R}^m_+$. Prove that the set

$$X = \bigcup_{\lambda \in \Lambda} \left(\sum_{i=1}^{m} \lambda_i X_i \right), \quad \lambda = (\lambda_1, \dots, \lambda_m),$$

is convex. Give an example that shows that the condition $\Lambda \subset \mathbb{R}^m_+$ is significant here.

18) Prove the Polterovich and Spivak theorem. Let X be a cone in $\mathbb{R}^n$, $X \subset \mathbb{R}^n_+$, and let, for all $x,\ y \in X$, the vectors

$$\min\{x,y\} = (\min\{x_1,y_1\},\ldots,\min\{x_n,y_n\}),$$

$$\max\{x,y\} = (\max\{x_1,y_1\},\ldots,\max\{x_n,y_n\})$$

belong to X. Then X is a convex cone. Give an example that shows that conditions of the theorem are essential here.

19) Let X be a closed and convex set, let $x^k \in X$, $\lambda_k \geq 0$ $k = 1, 2, \ldots$, $\sum\limits_{k=1}^{\infty} \lambda_k = 1$ and let the point $x = \sum\limits_{k=1}^{\infty} \lambda_k x^k$ exist, that is, the series converges. Prove that $x \in X$. Is the condition of closeness of X essential here?

20) a) Let X_1 be an arbitrary set, let X_2 be a convex set and let A be a bounded set in $\mathbb{R}^n$, such that $X_1 + A \subset X_2 + A$. Prove that $X_1 \subset \overline{X_2}$.

b) Let X_1 and X_2 be closed convex sets, and let A be a bounded set in $\mathbb{R}^n$, such that $X_1 + A = X_2 + A$. Prove that $X_1 = X_2$.

21) Find convex and conic hulls of the following sets in $\mathbb{R}^2$:

a) $X_1 = \left\{x \in \mathbb{R}^2 \,\middle|\, x_1^2 = x_2 \right\}$;

b) $X_2 = \left\{x \in \mathbb{R}^2 \,\middle|\, x_1^2 = x_2,\ x_1 \geq 0 \right\}$;

c) $X_3 = \left\{x \in \mathbb{R}^2 \,\middle|\, x_1 x_2 = 1 \right\}$;

d) $X_4 = \left\{x \in \mathbb{R}^2 \,\middle|\, \sin(x_1) = x_2,\ 0 \leq x_1 \leq \pi \right\}$;

e) $X_5 = \left\{x \in \mathbb{R}^2 \,\middle|\, e^{x_1} = x_2 \right\}$.

22) Prove that the convex hull of an open set is open.

23) Is it true that the convex hull of a closed set and the conic hull of a convex set are closed?

24) Let X be an arbitrary set in $\mathbb{R}^n$. Prove that

$$\operatorname{cone}(\operatorname{conv} X) = \operatorname{conv}(\operatorname{cone} X) = \operatorname{cone} X,$$

$$\operatorname{aff}(\operatorname{conv} X) = \operatorname{conv}(\operatorname{aff} X) = \operatorname{aff} X,$$

and if $0 \in X$, then

$$\operatorname{aff}(\operatorname{cone} X) = \operatorname{cone}(\operatorname{aff} X) = \operatorname{aff} X,$$

and in this case $\operatorname{aff} X$ is a linear subspace.

Is the last formula correct for arbitrary X?

25) Draw the sum of $X_1 + X_2$ of the following pairs of sets in $\mathbb{R}^2$:

$$X_1 = \text{conv}\ \{(1,2), (1,-3)\},$$

$$X_2 = \text{conv}\ \{(2,-3), (4,-1), (0,1), (3,2)\};$$

26) Draw the sum of $X_1 + X_2$ of the following pairs of sets in $\mathbb{R}^2$:

$$X_1 = \text{conv}\ \{(1,1), (-3,1), (2,-4)\},$$

$$X_2 = \text{conv}\ \{(-1,0), (1,3), (-2,2)\}.$$

27) Let $A\colon \mathbb{R}^n \to \mathbb{R}^m$ be a linear mapping, and let X be an arbitrary set in $\mathbb{R}^n$. Prove that

$$\text{conv}\, A(X) = A(\text{conv}\, X),$$

$$\text{cone}\, A(X) = A(\text{cone}\, X),$$

$$\text{aff}\, A(X) = A(\text{aff}\, X).$$

28) Let $X_1, \dots X_m$ be arbitrary sets in $\mathbb{R}^n$. Prove that

$$\text{aff}\left(\sum_{i=1}^m X_i\right) = \sum_{i=1}^m \text{aff}\, X_i,$$

and if $0 \in X_1, \dots, 0 \in X_m$, then

$$\text{cone}\left(\sum_{i=1}^m X_i\right) = \sum_{i=1}^m \text{cone} X_i.$$

Is the last equality true for arbitrary sets?

29) Let $X_1, \dots X_m$ be arbitrary sets in $\mathbb{R}^n$. Prove that

$$\text{cone}\left(\bigcup_{i=1}^m X_i\right) = \sum_{i=1}^m \text{cone}\, X_i,$$

and if $0 \in X_1, \dots, 0 \in X_m$, then

$$\text{aff}\left(\bigcup_{i=1}^m X_i\right) = \sum_{i=1}^m \text{aff}\, X_i.$$

Is the last equality true for arbitrary sets?

30) Let $X_i = \operatorname{conv} Y_i$, where $Y_i, i = 1, \ldots, n$ are arbitrary sets in $\mathbb{R}^n$. Prove that

$$\operatorname{conv}\left(\bigcup_{i=1}^{m} X_i\right) = \operatorname{conv}\left(\bigcup_{i=1}^{m} Y_i\right).$$

31) Let $X_i = \operatorname{conv} Y_i + \operatorname{cone} Z_i$, where $Y_i, Z_i, i = 1, \ldots, m$ are arbitrary sets in $\mathbb{R}^n$. Prove that

$$\overline{\operatorname{conv}\left(\bigcup_{i=1}^{m} X_i\right)} = \overline{\operatorname{conv}\left(\bigcup_{i=1}^{m} Y_i\right) + \operatorname{cone}\left(\bigcup_{i=1}^{m} Z_i\right)}.$$

32) Five points are given on the plane: $x^1 = (-2,\ 2)$; $x^2 = (4,\ 1)$; $x^3 = (1,\ 4)$; $x^4 = (-1,\ 3)$; $x^5 = (3,\ 3)$, as well as the point $x^0 = (2,\ 1)$, which belongs to $\operatorname{cone}\{x^1, \ldots, x^5\}$. Give all sets of numbers $\{i, j\} \subset \{1, \ldots, 5\}$ such that $x^0 \in \operatorname{cone}\{x^i, x^j\}$ to illustrate the fact that any point in $\operatorname{cone} X$, where $X \subset \mathbb{R}^n$ is an arbitrary set, can be represented as a non-negative combination of no more than n points from X.

33) On the plane there are given points: $x^1 = (-3, 1)$; $x^2 = (-3, 3)$; $x^3 = (2, 5)$; $x^4 = (4, 4)$; $x^5 = (5, -2)$; $x^6 = (0, -1)$, as well as the point $x^0 = (3, 1)$, which belongs to $\operatorname{conv}\{x^1, \ldots, x^6\}$. To illustrate the Carathéodory theorem, give all sets of numbers $\{i, j, k\} \subset \{1, \ldots, 6\}$, such that $x^0 \in \operatorname{conv}\{x^i, x^j, x^k\}$.

34) In the space $\mathbb{R}^3$ there are given points: $x^1 = (2, 8, 2)$, $x^2 = (6, 2, 2)$, $x^3 = (0, -8, 4)$, $x^4 = (9, -6, 3)$, $x^5 = (-6, 9, 3)$, as well as the point $x^0 = (1, 1, 1)$, which belongs to $\operatorname{conv}\{x^1, \ldots, x^5\}$. Give all sets of numbers $\{i, j, k\} \subset \{1, \ldots, 5\}$, such that $x^0 \in \operatorname{cone}\{x^i, x^j, x^k\}$.

35) In the space $\mathbb{R}^3$ there are given points: $x^1 = (1, -1, 1)$, $x^2 = (-5, 12, 7)$, $x^3 = (3, 1, 0)$, $x^4 = (-2, 3, 3)$, $x^5 = (2, 4, 9)$, $x^6 = (2, 3, 4)$, as well as the point $x^0 = (-1, 9, 16)$, which belongs to $\operatorname{cone}\{x^1, \ldots, x^6\}$. Represent x^0 as a non-negative combination of points $x^1, \ldots, x^6$ so that the sum of coefficients is minimal.

36) Points $x^1, \ldots, x^m \in \mathbb{R}^n$ are called *affinely dependent*, if there exist numbers $\lambda_1, \ldots, \lambda_m$, which are not all equal to zero and such that

$$\sum_{i=1}^{m} \lambda_i x^i = 0, \quad \sum_{i=1}^{m} \lambda_i = 1.$$

In the opposite case, points $x^1, \ldots, x^m \in \mathbb{R}^n$ are called *affinely independent.*

Prove that the following statements are equivalent:

a) points $x^1, \ldots, x^m$ are affinely independent;

b) points $x^2 - x^1, \ldots, x^m - x^1$ are linearly independent;

c) points $(x^1, 1), \ldots, (x^m, 1)$ are linearly independent.

From this property, it follows that the maximum number of affinely independent points in $\mathbb{R}^n$ is equal to $n+1$.

37) The convex hull of $m+1$ affinely independent points is called an m-simplex. Prove that

a) the dimension of the m-simplex is equal to m;

b) the dimension of a convex set X coincides with the maximal dimension of simplexes belonging to X.

38) If the point x is a convex (affine) combination of points $x^1, \dots, x^m$, then x can be presented as a convex (affine) combination of an affinely independent subsystem of these points. Prove this.

39) Let $x^1, \dots, x^m$ be linear independent points and let x be their linear combination, that is $x = \sum_{i=1}^{m} \lambda_i x^i$, and $\lambda_m \neq 0$. Prove that the points $x^1, \dots, x^{m-1}, x$ are linear independent.

40) Let $x^1, \dots, x^m$ be affinely independent points and let x be their affine combination, that is $x = \sum_{i=1}^{m} \lambda_i x^i, \quad \sum_{i=1}^{m} \lambda_i = 1$, and $\lambda_m \neq 0$. Prove that the points $x^1, \dots, x^{m-1}, x$ are affinely independent.

41) Prove that projection of any point $a \in \mathbb{R}^n$ on a closed convex set $X \subset \mathbb{R}^n$ is unique.

42) Prove the inverse of the statement in lemma 2.10. Let X be an arbitrary set of $\mathbb{R}^n$, and let $a \in \mathbb{R}^n$. If a point $\hat{x} \in X$ satisfies the condition $\langle a - \hat{x}, x - \hat{x} \rangle \leq 0$ for all $x \in X$, then $\hat{x}$ is a projection of the point a into the set X.

43) Let X be a closed convex set in $\mathbb{R}^n$. Prove that the projection operator $\pi_X(a)$ has the property of non-extension of distances, that is,

$$\left\|\pi_X(a^1) - \pi_X(a^2)\right\| \leq \left\|a^1 - a^2\right\| \text{ for all } a^1, a^2 \in \mathbb{R}^n.$$

44) Derive the equation of the hyperplane, which is supporting of the set

$$X = \left\{x \in \mathbb{R}^3 \left| x_1^2 - 2x_1x_2 + 10x_2^2 + 6x_2x_3 + x_3^3 \leq 25 \right.\right\}$$

at point $x^0 = (4, 1, 1)$.

45) Derive the equation of the hyperplane, which is supporting of the set

$$X = \left\{x \in \mathbb{R}^3 \left| x_3 \geq x_1^2 + x_2^2 \right.\right\},$$

and separates it from the point $x^0 = (-5/4, 5/16, 15/16)$.

46) Find out for which values of the parameter k the hyperplane $H_{p\beta}$, where $p = (-3k, 12, 2k)$ and $\beta = 12$, is supporting at the point $x^0 = (2,1,3)$ to the polyhedron $X \subset \mathbb{R}^3$, given by the system of inequalities

$$\begin{cases} x_1 + 8x_2 + x_3 \leq 13, \\ -2x_1 + 3x_2 + x_3 \leq 2, \\ 3x_1 - 2x_2 - x_3 \leq 2, \\ -5x_1 + x_2 + 2x_3 \leq -3. \end{cases}.$$

47) Find all values of the parameter k, for which the hyperplane $H_{p\beta}$, where $p = (k, 1, -2, -k^2 - 4k + 4, -1)$ and $\beta = 2k + 1$, is supporting at the point $x^0 = (2, 1, 0, 0, 0)$ to the polyhedron $X \subset \mathbb{R}^5$, given by the system

$$\begin{cases} x_1 - x_2 + 2x_3 - 3x_4 = 1, \\ 2x_1 - x_2 + 3x_3 - 5x_4 + x_5 = 3, \end{cases},$$

$$x_j \geq 0, \quad j = 1, \ldots, 5.$$

48) Derive the equation of a hyperplane separating the convex sets

$$X_1 = \left\{ x \in \mathbb{R}^2 \,\middle|\, \frac{x_1^2}{4} + \frac{x_2^2}{9} \leq 1 \right\},$$

$$X_2 = \left\{ x \in \mathbb{R}^2 \,\middle|\, x_2 \geq \frac{3}{x_1},\ x_1 > 0 \right\}.$$

49) Suppose that a hyperplane $H_{p\beta}$ is supporting a set X at a point $a \in \overline{X}$, and the point a is a convex positive combination of the points $x^1, \ldots, x^m$ from $\overline{X}$, that is, $a = \sum_{i=1}^{m} \lambda_i x^i$, $\lambda_i > 0$, $i = 1, \ldots, m$, $\sum_{i=1}^{m} \lambda_i = 1$. Prove that these points lie in $H_{p\beta}$.

50) Prove that every hyperplane supporting a cone passes through zero.

51) Prove that every support hyperplane supporting an affine set contains this affine set.

52) Prove that a hyperplane $H_{p\beta}$ supporting a ball $U_\varepsilon(x^0)$ at a boundary point a is unique and is determined by the parameters $p = a - x^0$, $\beta = \langle p, a \rangle$.

53) Let X be a bounded set in $\mathbb{R}^n$. Prove that for arbitrary $p \in \mathbb{R}^n$, there exists a number β such that the hyperplane $H_{p\beta}$ is supporting X.

54) Let a polyhedron X lie in the half-space $H_{p\beta}^-$. Prove that there exists a number $\alpha \leq \beta$ such that the hyperplane $H_{p\beta}$ is supporting X.

55) A closed set $X \subset \mathbb{R}^n$ is called strictly convex if $\lambda x^1 + (1-\lambda)\, x^2 \in \operatorname{int} X$ for all $x^1, x^2 \in X$, $x^1 \neq x^2$, $\lambda \in (0,1)$. Show that the ball $U_\varepsilon(x^0)$ is a strictly convex set, and the cube

$$K_\varepsilon(x^0) = \left\{x \in \mathbb{R}^n \colon \left|x_j - x_j^0\right| \leq \varepsilon, j = 1, \ldots, n\right\}$$

is not a strictly convex set.

56) Let X be a closed convex set in $\mathbb{R}^n$. Prove that X is strictly convex if and only if every support to X at a point $a \in \partial X$ hyperplane $H_{p\beta}$ intersects with X only at this point: $H_{p\beta} \cap X = \{a\}$.

57) Prove such a refinement of the Carathéodory theorem. Let X be a set in $\mathbb{R}^n$. Then every point of conv X, which lies on the boundary of conv X, can be represented as a convex combination of not more than n points from X.

58) Let X be a closed convex set in $\mathbb{R}^n$, and $X \cap \mathbb{R}^n_+ = \{0\}$. Does it follow from this that there exists a vector $p > 0$ for which $\langle p, x\rangle \leq 0$ for all $x \in X$?

59) Let X be a closed convex set in $\mathbb{R}^n$ different from $\mathbb{R}^n$, and let its complement $\mathbb{R}^n \backslash X$ also be convex. Prove that X is a half-space.

60) Let X_1 and X_2 be convex sets in $\mathbb{R}^n$, with $\operatorname{ri} X_1 \cap \operatorname{ri} X_2 = \emptyset$. Suppose that at least one of the conditions is fulfilled additionally: (a) X_1 is bounded; (b) X_1 is a cone. Prove that X_1 and X_2 can be properly separated by means of a hyperplane supporting X_1.

61) Prove that the polyhedra X_1 and X_2 from $\mathbb{R}^n$ that do not intersect are strongly separated.

62) *Voronoi sets*. Let $x^0, x^1, \ldots, x^K \in \mathbb{R}^n$. Define a set of points that are closer (in the Euclidean norm) to x^0 than to $x^i, i = 1, 2, \ldots, K$, that is,

$$V = \{x \in \mathbb{R}^n \,|\, \|x - x^0\|_2 \leq \|x - x^i\|_2,\ i = 1, \ldots, K\}.$$

The set V is called the *Voronoi set* around x^0 related to points $x^i, i = 1, 2, \ldots, K$.

1) Show that the set V is a polyhedron. Represent V in the form

$$X = \{x \in \mathbb{R}^n \colon Ax \leq b\}.$$

2) On the other hand, for a given polyhedron with a non-empty set of interior points, find points $x^0, x^1, \ldots, x^K \in \mathbb{R}^n$ such that the given polyhedron is the Voronoi set around x^0 in relation to $x^i, i = 1, 2, \ldots, K$.

3) Determine a set of points

$$V_k = \{x \in \mathbb{R}^n \,|\, \|x - x^k\|_2 \leq \|x - x^i\|_2,\ i = 1, \ldots, K;\ i \neq k\}.$$

The sets $V_0, V_1, \ldots, V_K$ determine a polyhedral decomposition of the space $\mathbb{R}^n$, that is, the set $V_0, V_1, \ldots, V_K$ are polyhedra,

$$\cup_{k=0}^K V_k = \mathbb{R}^n, \quad \operatorname{int} V_i \cap \operatorname{int} V_j = \emptyset, i \neq j.$$

Let $P_1, \ldots, P_M$ be polyhedra such that

$$\cup_{k=1}^{M} P_k = \mathbb{R}^n, \quad \operatorname{int} P_i \cap \operatorname{int} P_j = \emptyset, i \neq j.$$

Is it possible to determine such a polyhedral decomposition of the space $\mathbb{R}^n$ as a decomposition of the Voronoi sets with an appropriate choice of points $x^0, x^1, \ldots, x^K \in \mathbb{R}^n$?

63) Let $X_1, \ldots, X_m$ be convex sets in $\mathbb{R}^n$, with $\bigcap_{i=1}^{m} X_i = \emptyset$. Prove that these sets are separated.

The sets $X_1, \ldots, X_m$ in $\mathbb{R}^n$ are

a) separated, if there are exist vectors $p^1, \ldots, p^m \in \mathbb{R}^n$ that are not all equal to zero and numbers $\beta_1, \ldots, \beta_m \in \mathbb{R}^1$, such that

$$\langle p^i, x^i \rangle \leq \beta_i \quad \forall x^i \in X_i, i = 1, \ldots, m; \quad \sum_{i=1}^{m} p^i = 0, \quad \sum_{i=1}^{m} \beta_i \leq 0;$$

b) properly separated, if, in addition, $\langle p^i, \bar{x}^i \rangle < \beta_i$ for some i and $\bar{x}^i \in X_i, i = 1, \ldots, m$;

c) strongly separated, if the separation conditions are satisfied and $\sum_{i=1}^{m} \beta_i < 0$.

3

Convex Functions

3.1. Convex functions: basic definitions

DEFINITION 3.1.– *A function* $f\colon \mathbb{R}^n \to \mathbb{R}$ *defined on a convex set* $X \subset \mathbb{R}^n$ *is said to be convex, if the following inequality holds true:*

$$f\left(\lambda x^1 + (1-\lambda)\, x^2\right) \leqslant \lambda f\left(x^1\right) + (1-\lambda)\, f\left(x^2\right) \qquad [3.1]$$

for all $x^1, x^2 \in X$*, and all* $\lambda \in [0,1]$.

If for all $x^1, x^2 \in X$, $x^1 \neq x^2$*, and all* $\lambda \in (0,1)$ *the strict inequality holds true, then the function* f *is said to be strictly convex on the set* X.

We define another important subclass of convex functions.

DEFINITION 3.2.– *A function* $f\colon \mathbb{R}^n \to \mathbb{R}$ *defined on a convex set* $X \subset \mathbb{R}^n$ *is said to be* strongly convex *on* X*, if there exists a constant* $\theta > 0$ *such that the following inequality holds true:*

$$f\left(\lambda x^1 + (1-\lambda)\, x^2\right) \leqslant \lambda f\left(x^1\right) + (1-\lambda)\, f\left(x^2\right) - \theta\lambda\,(1-\lambda)\left\|x^1 - x^2\right\|^2 \qquad [3.2]$$

for all $x^1, x^2 \in X$*, and all* $\lambda \in [0,1]$*. The constant* $\theta > 0$ *is called* the module of strong convexity *of the function* $f(x)$.

A simple example of a strongly convex function is the function $f(x) = \|x\|^2$ on $\mathbb{R}^n$. For this function, inequality [3.2] holds true as equality with $\theta = 1$.

A strongly convex function is, obviously, strictly convex. The reverse statement is not true. For example, the function $f(x) = x^4$ (geometrically "similar" to strongly convex function $f(x) = x^2$) is strictly, but not strongly, convex on $\mathbb{R}$ (see formula [3.16]).

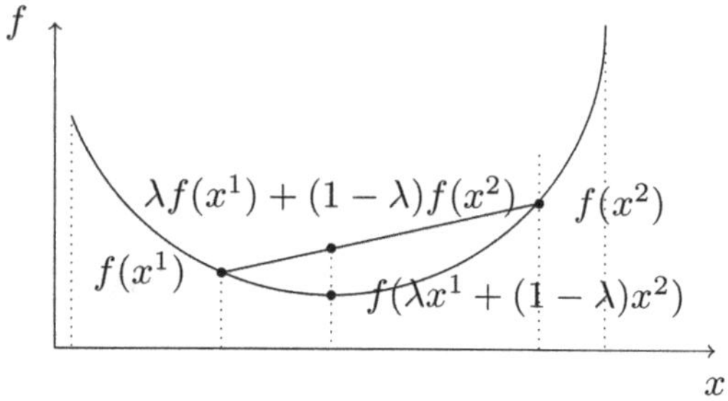

Figure 3.1. *Convex function*

The following theorem describes a connection between functions from the class of convex functions and functions from the class of strongly convex functions.

THEOREM 3.1.– A function f is strongly convex with the modulus of convexity $\theta > 0$ on a convex set $X \subset \mathbb{R}^n$ if and only if the function $f(x) - \theta \|x\|^2$ is convex on X.

PROOF.– To prove this theorem, we use the definition of convexity of the function $f(x) - \theta \|x\|^2$:

$$f\left(\lambda x^1 + (1-\lambda) x^2\right) - \theta \left\|\lambda x^1 + (1-\lambda) x^2\right\|^2 \leqslant$$
$$\leqslant \lambda f\left(x^1\right) + (1-\lambda) f\left(x^2\right) - \theta \left[\lambda \left\|x^1\right\|^2 + (1-\lambda) \left\|x^2\right\|^2\right]$$

and equality [3.2] with $\theta = 1$ for strongly convex function $f(x) = \|x\|^2$ on $\mathbb{R}^n$. □

This simple theorem shows that properties of convex functions can be studied by examining the corresponding properties of strongly convex functions.

DEFINITION 3.3.– *A function $f\colon \mathbb{R}^n \to \mathbb{R}$ defined on a convex set $X \subset \mathbb{R}^n$ is called (strictly, strongly) concave on X if the function $g = -f$ is (strictly, strongly) convex on X.*

EXAMPLE 3.1.– It is easy to check the convexity properties of the following functions:

– $f(x) = e^{ax}$ is convex on $\mathbb{R}$;

– $f(x) = x^a$ is convex on $\operatorname{int} \mathbb{R}_+$ when $a \geqslant 1$ or $a \leqslant 0$, and it is concave when $0 \leqslant a \leqslant 1$;

– $f(x) = |x|^p$ with $p \geqslant 1$ is convex on $\mathbb{R}$.

EXAMPLE 3.2.– *Convexity of the norm.* Every norm on $\mathbb{R}^n$ is convex.

If f: $\mathbb{R}^n \to \mathbb{R}$ is a norm on $\mathbb{R}^n$ and $0 \leqslant \lambda \leqslant 1$, then

$$f(\lambda x + (1-\lambda)y) \leqslant f(\lambda x) + f((1-\lambda)x) = \lambda f(x) + (1-\lambda)f(y),$$

since by definition the norm is a homogeneous function and satisfies the triangle inequality.

EXAMPLE 3.3.– *Convexity of the maximum.* The function

$$f(x) = \max\{x_1, \ldots, x_n\}$$

is convex on $\mathbb{R}^n$. It satisfies inequality for $0 \leqslant \lambda \leqslant 1$

$$f(\lambda x + (1-\lambda)y) = \max_{1\leqslant i\leqslant n} \{\lambda x_i + (1-\lambda)y_i\}$$

$$\leqslant \lambda \max_{1\leqslant i\leqslant n} \{x_i\} + (1-\lambda) \max_{1\leqslant i\leqslant n} \{y_i\} =$$

$$= \lambda f(x) + (1-\lambda)f(y).$$

Inequality [3.1], which defines a convex function, can be extended for an arbitrarily finite number of points.

THEOREM 3.2.– Let f be a convex function on a convex set X. Then

$$f\left(\sum_{i=1}^{m} \lambda_i x^i\right) \leqslant \sum_{i=1}^{m} \lambda_i f\left(x^i\right), \qquad [3.3]$$

for all $m = 1, 2 \ldots$; $x^i \in X$, $\lambda_i \geqslant 0$, $i = 1, \ldots, m$, $\sum_{i=1}^{m} \lambda_i = 1$.

PROOF.– Apply the induction on m. If $m = 1$, then inequality [3.3] is obvious. Let this be proved for $m = k$. Let us prove it for $m = k + 1$. Let $x = \sum_{i=1}^{k+1} \lambda_i x^i$, where (we reject the trivial case) we will assume that $\lambda_{k+1} < 1$. Then we can use [3.1]. Using first the convexity of f, and then the induction, we have

$$f(x) \leqslant (1-\lambda_{k+1}) f\left(\sum_{i=1}^{k} \frac{\lambda_i}{1-\lambda_{k+1}} x^i\right) + \lambda_{k+1} f\left(x^{k+1}\right)$$

$$\leqslant (1-\lambda_{k+1}) \sum_{i=1}^{k} \frac{\lambda_i}{1-\lambda_{k+1}} f\left(x^i\right) + \lambda_{k+1} f\left(x^{k+1}\right) = \sum_{i=1}^{k+1} \lambda_i f\left(x^i\right).$$

□

Relation [3.3] is the famous *Jensen inequality*. It can be used to prove a number of known inequalities. Let us restrict ourselves to the following example.

EXAMPLE 3.4.– The function $f(x) = -\ln x$ is convex on $\operatorname{int}\mathbb{R}_+$ (see formula [3.16]). For this reason for all $m = 1, 2, \ldots;\ x_i > 0,\ \lambda_i \geqslant 0,\ i = 1, \ldots, m$, $\sum_{i=1}^{m} \lambda_i = 1$ we have

$$-\ln\left(\sum_{i=1}^{m} \lambda_i x_i\right) \leqslant -\sum_{i=1}^{m} \lambda_i \ln x_i = -\ln\left(\prod_{i=1}^{m} x_i^{\lambda_i}\right).$$

From this inequality, we have

$$\sum_{i=1}^{m} \lambda_i x_i \geqslant \prod_{i=1}^{m} x_i^{\lambda_i}.$$

In the particular case where $\lambda_i = 1/m, i = 1, 2, ..., m$, we obtain the classical inequality between the arithmetic mean and the geometric mean

$$\frac{1}{m}\sum_{i=1}^{m} x_i \geqslant \left(\prod_{i=1}^{m} x_i\right)^{\frac{1}{m}}.$$

The definition of (strictly, strongly) convex functions involves functions defined on a set X, which must be convex. A more general definition of convex functions can be given for functions defined on the entire space $\mathbb{R}^n$ that can take infinite values.

DEFINITION 3.4.– *The function* $f\colon \mathbb{R}^n \to \mathbb{R} \cup \{+\infty\}$, *which is not equal identically to* $+\infty$, *is called convex, if for all* $x^1, x^2 \in \mathbb{R}^n$, *and all* $\lambda \in (0, 1)$ *the following inequality holds*

$$f\left(\lambda x^1 + (1-\lambda)\, x^2\right) \leqslant \lambda f\left(x^1\right) + (1-\lambda)\, f\left(x^2\right),$$

which is considered as an inequality in $\mathbb{R} \cup \{+\infty\}$. *We denote the set of such functions by* Conv $\mathbb{R}^n$.

DEFINITION 3.5.– *The* effective set *of a function* $f \in \operatorname{Conv} \mathbb{R}^n$ *is the non-empty set*

$$\operatorname{dom} f\colon = \{x \in \mathbb{R}^n\colon f(x) < +\infty\}.$$

To each convex function f on a convex set X, we can define the corresponding function

$$\tilde{f}(x) = \begin{cases} f(x), & x \in X; \\ +\infty, & x \notin X \end{cases}$$

from the class Conv $\mathbb{R}^n$. And vice versa, for $f \in$ Conv $\mathbb{R}^n$ we can take $X :=$ dom f to obtain a convex function on a convex set X. The same correspondence is between strictly (strongly) convex functions on a convex set X and strictly (strongly) convex functions $f\colon \mathbb{R}^n \to \mathbb{R} \cup \{+\infty\}$.

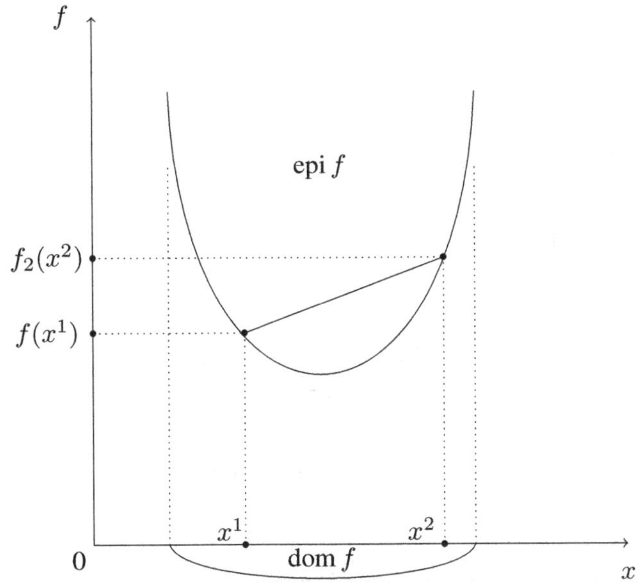

Figure 3.2. *Epigraph of convex function*

DEFINITION 3.6.– *The* epigraph *of a function $f\colon \mathbb{R}^n \to \mathbb{R} \cup \{+\infty\}$, which does not identically equal $+\infty$, is the non-empty set*

$$\text{epi } f :=\{(x, r) \in \mathbb{R}^n \times \mathbb{R} \colon f(x) \leq r\}.$$

The strict epigraph *of a function $f\colon \mathbb{R}^n \to \mathbb{R} \cup \{+\infty\}$, which does not identically equal $+\infty$, is the non-empty set*

$$\text{epi } f :=\{(x, r) \in \mathbb{R}^n \times \mathbb{R} \colon f(x) < r\}.$$

THEOREM 3.3.– Suppose that a function $f\colon \mathbb{R}^n \to \mathbb{R} \cup \{+\infty\}$ is not identically equal to$+\infty$. Then the following properties of the function f are equivalent:

i) $f \in$ Conv $\mathbb{R}^n$;

ii) the epigraph of the function f is a convex set in $\mathbb{R}^n \times \mathbb{R}$;

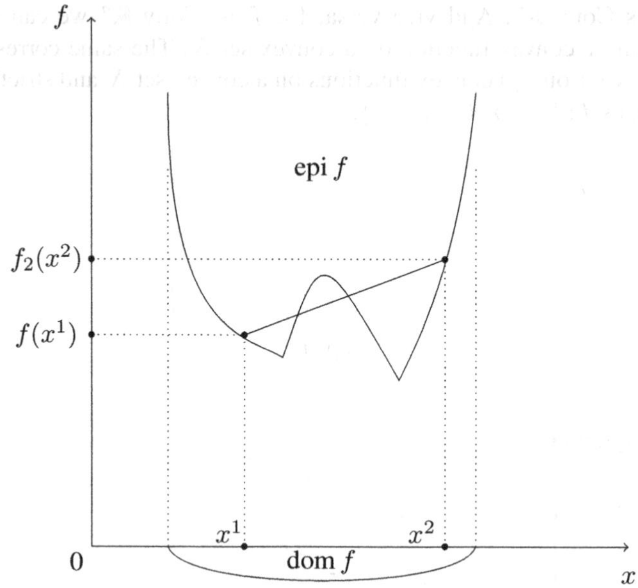

Figure 3.3. *Epigraph of nonconvex function*

iii) the strict epigraph of the function f is a convex set in $\mathbb{R}^n \times \mathbb{R}$.

DEFINITION 3.7.– *The* sublevel set *of a function* $f: \mathbb{R}^n \to \mathbb{R} \cup \{+\infty\}$*, which is not equal identically to* $+\infty$*, is a non-empty set*

$$S_r(f) = \{x \in \mathbb{R}^n : f(x) \leqslant r\}, \quad r \in \mathbb{R}.$$

This set is also called the Lebesgue set *of a function* f.

It follows from the definition that the following property holds true

$$(x, r) \in \text{epi } f \Leftrightarrow x \in S_r(f).$$

Every sublevel set of a function $f: \mathbb{R}^n \to \mathbb{R} \cup \{+\infty\}$ is convex. We can construct the set $S_r(f)$ with the help of the intersection of epigraph epi f of the function f with the horizontal hyperplane $\mathbb{R}^n \times \{r\}$, and, next, construct projection onto $\mathbb{R}^n \times \{0\}$ of the intersection $(\text{epi } f) \cap (\mathbb{R}^n \times \{r\})$ of two convex sets. This operation changes the topology. However, it does not really change the set of relative interior points. The following theorem is true.

THEOREM 3.4.– Let a function $f\colon \mathbb{R}^n \to \mathbb{R} \cup \{+\infty\}$. The set of relative interior points, ri epi f, is a union of rays with a base at points $f(x)$, $x \in$ ri dom f:

$$\text{ri epi } f = \{(x, r) \in \mathbb{R}^n \times \mathbb{R} : x \in \text{ri dom } f, r > f(x)\}.$$

PROOF.– Since dom f is a projection of epi f on $\mathbb{R}^n$ (the operation of making projection is linear), and the set ri dom f is a projection of ri epi f on $\mathbb{R}^n$. □

EXAMPLE 3.5.– The epigraph of a linear function $f(x) = \langle s, x\rangle$,

$$\text{epi } f = \{(x, r) \in \mathbb{R}^n \times \mathbb{R} | r \geq \langle s, x\rangle\},$$

is a closed half-space determined by a vector $s \in \mathbb{R}^n$. The epigraph of an *affine function* $f(x) = \langle s, x\rangle + b$ can be represented in terms of a point $x_0 \in \mathbb{R}^n$ in the form

$$\text{epi } f = \{(x, r) \in \mathbb{R}^n \times \mathbb{R} | r \geq f(x_0) + \langle s, x - x_0\rangle\}$$

$$= \{(x, r) \in \mathbb{R}^n \times \mathbb{R} | \langle s, x\rangle - r \leq \langle s, x_0\rangle - f(x_0)\}.$$

This is a closed half-space determined by the vector $(s, -1) \in \mathbb{R}^n \times \mathbb{R}$ and a certain constant.

THEOREM 3.5.– For each function $f \in \text{Conv } \mathbb{R}^n$, there exists an affine function, such that

$$f(x) \geq f(x_0) + \langle s, x - x_0\rangle \quad \forall x \in \mathbb{R}^n.$$

This means that each function $f \in \text{Conv } \mathbb{R}^n$ has a support affine function.

PROOF.– The effective set, dom f, of a function f is an image of the epigraph epi f (a result of projection of epi f on $\mathbb{R}^n$ which is a linear operation). That is why aff epi f = (aff dom f) $\times \mathbb{R}$. Denote by V the linear subpace parallel to aff dom f. Then aff dom $f = V + \{x_0\}$, where x_0 is an arbitrary point from dom f. We have

$$\text{aff epi } f = (V + \{x_0\}) \times \mathbb{R}.$$

Let $x_0 \in$ ri dom f. It follows from the previous theorem that $(x_0, f(x_0)) \in \partial(\text{epi } f)$ and we can construct a non-trivial hyperplane, which is a support hyperplane to epi f at point $(x_0, f(x_0))$. Therefore, there exist $s = s_v \in V$ and $\alpha \in \mathbb{R}$ such that

$$\langle s, x\rangle + \alpha r \leq \langle s, x_0\rangle + \alpha f(x_0)$$

for all $(x, r) : f(x) \leq r$. For this reason, we have $\alpha \leq 0$ ($r \to \infty$). We can choose $s \in V$ and $x_0 \in$ ri dom f and find a small δ, such that $x_0 + \delta s \in$ dom f and

$$\delta \|s\|^2 \leq \alpha [f(x_0) - f(x_0 + \delta s)] < +\infty.$$

That is why $\alpha \neq 0$. We can now take $\alpha = -1$ and get the statement of the theorem. □

The theorem states that each convex epigraph has a supporting non-vertical hyperplane. As a result, each convex function is bounded from below on every bounded set in $\mathbb{R}^n$.

DEFINITION 3.8.– *A convex function* $f\colon \mathbb{R}^n \to \mathbb{R} \cup \{+\infty\}$ *is called* closed *if its epigraph,* epi f, *is a closed set in* $\mathbb{R}^n \times \mathbb{R}$. *We will denote the set of such functions by* $\overline{\text{Conv}}\ \mathbb{R}^n$.

THEOREM 3.6.– Let a function $f\colon \mathbb{R}^n \to \mathbb{R} \cup \{+\infty\}$. The following properties of the function f are equivalent:

i) the function f is lower semicontinuous on $\mathbb{R}^n$;

ii) the function f is closed;

iii) the sublevel set $S_r(f) = \{x \in \mathbb{R}^n : f(x) \leqslant r\}$ of the function f is closed in $\mathbb{R}^n$ for all $r \in \mathbb{R}$.

PROOF.– $(i) \Rightarrow (ii)$. Let $(y_k, r_k), k = 1, 2, \ldots$ be a set of points from epi f that converges to (x, r) as $k \to \infty$. Since $f(y_k) \leq r_k$ for all k, then

$$r = \lim_{k\to\infty} r_k \geq \liminf_{k\to\infty} f(y_k) \geq \liminf_{y\to x} f(y) \geq f(x).$$

This means that $(x, r) \in \text{epi } f$.

(ii) $\Rightarrow$ (iii). The intersection $\{\text{epi } f\} \cap \{\mathbb{R}^n \times \{r\}\}$ of two convex sets is a convex set.

(iii) $\Leftrightarrow$ (i). Theorem 1.8. □

DEFINITION 3.9.– *The* closure (lower semicontinuous hull) *of a function* $f\colon \mathbb{R}^n \to \mathbb{R} \cup \{+\infty\}$ *is determined by the relation*

$$\text{cl } f(x) = \liminf_{y\to x} f(y) \quad \forall x \in \mathbb{R}^n,$$

or (equivalent)

$$\text{epi}\,(\text{cl } f)\colon = \text{cl}\,(\text{epi } f).$$

THEOREM 3.7.– The closure of a function $f \in \text{Conv } \mathbb{R}^n$ can be represented as a supremum of the support to f affine functions

$$\text{cl } f(x) = \sup_{(s,b)\in\mathbb{R}^n\times\mathbb{R}} \{\langle s, x\rangle - b | \langle s, y\rangle - b \leq f(y)\, \forall\, y \in \mathbb{R}^n\}. \qquad [3.4]$$

PROOF.– A closed subspace that contains epi f is determined by a non-trivial vector $(s,\alpha) \in \mathbb{R}^n \times \mathbb{R}$ and a number b such that

$$\langle s, x\rangle + \alpha r \leq b \quad \forall (x,r) \in \text{epi } f.$$

Denote by $\Sigma \subset \mathbb{R}^n \times \mathbb{R} \times \mathbb{R}$ the set of all such indices $\sigma = (s,\alpha,b)$. Let

$$H_\sigma^- : = \{(x,r) | (x,r) + \alpha r \leq b\}.$$

denote the corresponding half-space. Then

$$\text{epi}(\text{cl } f) : = \text{cl}(\text{epi } f) = \cap_{\sigma\in\Sigma} H_\sigma^-.$$

It follows from the construction of the epigraph that in the inequality

$$\langle s, x\rangle + \alpha r \leq b \quad \forall (x,r) \in \text{epi } f$$

only $\alpha \leq 0$ is possible. Due to homogeneity, it is enough to consider cases $\alpha = 0$ and $\alpha = -1$. Let the set of indices Σ_1 correspond to $\alpha = -1$, and the set of indices Σ_0 correspond to $\alpha = 0$. The set Σ_1 determines a support to the function f affine sets. That is why $\Sigma_1 \neq \emptyset$. The set Σ_0 determines the closed subspaces in $\mathbb{R}^n$, which contain dom f ($\Sigma_0 = \emptyset$ when dom $f = \mathbb{R}^n$). Consider arbitrary $\sigma_0 = (s_0, 0, b_0) \in \Sigma_0$ and $\sigma_1 = (s_1, -1, b_1) \in \Sigma_1$ and construct

$$\sigma(t) : = (s_1 + ts_0, -1, b_1 + tb_0) \in \Sigma_1,\ t \geq 0.$$

Show that

$$H_{\sigma_0}^- \cap H_{\sigma_1}^- = \cap_{t\geq 0} H_{\sigma(t)}^- : = H^-.$$

If $(x,r) \in H_{\sigma_0}^- \cap H_{\sigma_1}^-$, then

$$\langle s_1 + ts_0, x\rangle - (b_1 + tb_0) \leq r \quad \forall t \geq 0,$$

which means $(x,r) \in H^-$. Now take $(x,r) \in H^-$. Then $(x,r) \in H_{\sigma_1}^-$ when $t = 0$. If $t > 0$, then as $t \to \infty$ we have $(x,r) \in H_{\sigma_0}^-$. □

3.2. Operations in the class of convex functions

In this section, we describe some operations on convex functions, results of which are also convex functions.

THEOREM 3.8.– Let $f_1, \ldots, f_m$ be convex functions on a convex set X, and let $\alpha_1, \ldots, \alpha_m$ be non-negative numbers. Then the function

$$f(x) = \sum_{i=1}^{m} \alpha_i f_i(x)$$

is convex on the set X. The function $f(x) = \sum_{i=1}^{m} \alpha_i f_i(x)$ is strictly (strongly) convex on X if at least one function f_i is strictly (strongly) convex and $\alpha_i > 0$.

PROOF.– For arbitrary $x^1, x^2 \in X, \lambda \in [0,1]$, we have

$$f\left(\lambda x^1 + (1-\lambda)x^2\right) = \sum\nolimits_{i=1}^{m} \alpha_i f_i\left(\lambda x^1 + (1-\lambda)x^2\right)$$

$$\leqslant \sum\nolimits_{i=1}^{m} \alpha_i \left(\lambda f_i\left(x^1\right) + (1-\lambda) f_i\left(x^2\right)\right) = \lambda f\left(x^1\right) + (1-\lambda) f\left(x^2\right).$$

This means that inequality [3.1] holds true. So the function f is convex. □

THEOREM 3.9.– Let X be a convex set, let Y be an arbitrary set and let $\varphi(x,y)$ be a function on $X \times Y$, which is convex with respect to x on X for every $y \in Y$ and bounded from above with respect to y on Y for every $x \in X$. The function

$$f(x) = \sup_{y \in Y} \varphi(x,y)$$

is convex on X.

PROOF.– For all $x^1, x^2 \in X, \lambda \in [0,1]$, we have

$$f\left(\lambda x^1 + (1-\lambda)x^2\right) = \sup_{y \in Y} \varphi\left(\lambda x^1 + (1-\lambda)x^2, y\right)$$

$$\leqslant \sup_{y \in Y} \left(\lambda \varphi\left(x^1, y\right) + (1-\lambda)\varphi\left(x^2, y\right)\right)$$

$$\leqslant \sup_{y \in Y} \left(\lambda \varphi\left(x^1, y\right)\right) + \sup_{y \in Y} \left((1-\lambda)\varphi\left(x^2, y\right)\right) = \lambda f\left(x^1\right) + (1-\lambda) f\left(x^2\right).$$

□

COROLLARY 3.1.– The function

$$f(x) = \max_{i=1,\ldots,m} \{f_i(x)\}$$

is convex on X if the functions $f_1(x), \ldots, f_m(x)$ are convex on X.

The key property is that a maximum of functions corresponds to an intersection of epigraphs:

$$\text{epi } f = \cap_{i=1,\ldots,m}(\text{epi } f_i)$$

which conceives convexity and closeness.

THEOREM 3.10.– Let X and Y be convex sets, and let $\varphi(x,y)$ be a function on $X \times Y$, which is jointly convex in $x \in X$ and $y \in Y$. Then the function

$$f(x) = \inf_{y \in Y} \varphi(x,y)$$

is convex on X provided that $f(x) > -\infty$ for all x.

PROOF.– The domain of f is the projection of dom f on its x-coordinates, i.e.

$$\operatorname{dom} f = \{x | (x, y) \in \operatorname{dom} \varphi \text{ for some } y \in Y\}.$$

We prove the convexity of f by verifying Jensen's inequality for $x^1, x^2 \in \operatorname{dom} f$. Let $\varepsilon > 0$. Then there are $y^1, y^2 \in Y$ such that $\varphi(x^i, y^i) \leq f(x^i) + \varepsilon$ for $i = 1, 2$. Let $\lambda \in [0, 1]$. We have

$$f(\lambda x^1 + (1-\lambda)x^2) = \inf_{y \in Y} \varphi(\lambda x^1 + (1-\lambda)x^2, y)$$

$$\leq \varphi(\lambda x^1 + (1-\lambda)x^2, \lambda y^1 + (1-\lambda)y^2) \leq \lambda\varphi(x^1, y^1) + (1-\lambda)\varphi(x^2, y^2)$$
$$\leq \lambda f(x^1) + (1-\lambda) f(x^2) + \varepsilon.$$

Since this holds for any $\varepsilon > 0$, we have

$$f(\lambda x^1 + (1-\lambda)x^2) \leq \lambda f(x^1) + (1-\lambda) f(x^2)$$

for all $\lambda \in [0, 1]$. □

THEOREM 3.11.– Let $g_1, \ldots, g_m$ be convex on the set $X \subset \mathbb{R}^n$ functions, let $g = (g_1, \ldots, g_m)$ be the corresponding vector function and let φ be a (coordinatewise) monotonically non-decreasing convex function on a convex set $U \subset \mathbb{R}^m$, where $g(X) \subset U$. The function

$$f(x) = \varphi(g(x))$$

is convex on X.

PROOF.– For all $x^1, x^2 \in X$, $\lambda \in [0, 1]$, we have

$$f\left(\lambda x^1 + (1-\lambda) x^2\right) = \varphi\left(g\left(\lambda x^1 + (1-\lambda) x^2\right)\right)$$

$$\leqslant \varphi\left(\lambda g\left(x^1\right) + (1-\lambda) g\left(x^2\right)\right) \leqslant \lambda\varphi\left(g\left(x^1\right)\right) + (1-\lambda)\varphi\left(g\left(x^2\right)\right)$$
$$= \lambda f\left(x^1\right) + (1-\lambda) f\left(x^2\right),$$

where the first inequality follows from the convexity of the function g and, monotonically non-decreasing, of the function φ, and the second inequality follows from the convexity of the function φ. □

THEOREM 3.12.– Let φ be a convex function on a convex set $U \subset \mathbb{R}^m$, let A be an $m \times n$ matrix, let $b \in \mathbb{R}^m$ and let the set $X = \{x \in \mathbb{R}^n \,| Ax + b \in U\}$ be non-empty. The function

$$f(x) = \varphi(Ax + b)$$

is convex on the set X.

PROOF.– For all $x^1, x^2 \in X, \lambda \in [0,1]$, we have

$$f\left(\lambda x^1 + (1-\lambda)x^2\right) = \varphi\left(A\left(\lambda x^1 + (1-\lambda)x^2\right) + b\right)$$
$$= \varphi\left(\lambda\left(Ax^1 + b\right) + (1-\lambda)\left(Ax^2 + b\right)\right)$$
$$\leqslant \lambda\varphi\left(Ax^1 + b\right) + (1-\lambda)\varphi\left(Ax^2 + b\right) = \lambda f\left(x^1\right) + (1-\lambda) f\left(x^2\right).$$

□

DEFINITION 3.10.– *Let f_1 and f_2 be convex functions. The function*

$$(f_1 \oplus f_2)(x) = \inf\{f_1(x_1) + f_2(x_2) \colon x_1 + x_2 = x\} \qquad [3.5]$$

is called the infimal convolution *of functions f_1, f_2.*

THEOREM 3.13.– Let $f_1, \ldots, f_m$ be convex functions on a convex set X. Then the function

$$(f_1 \oplus \cdots \oplus f_m)(x) = \inf\left\{\sum_{i=1}^{m} f_i(x_i) \colon \sum_{i=1}^{m} x_i = x\right\},$$

the infimal convolution of functions $f_1, \ldots, f_m$, is convex on the set X.

To prove that the infimal convolution of convex functions f_1, f_2 is convex, one can also show the following relation between strict epigraphs

$$\text{epi}\ (f_1 \oplus f_2) = \text{epi}\ f_1 + \text{epi}\ f_2.$$

That is why the infimal convolution of convex functions is sometimes called (strict) *epigraph addition.*

We can give an economic interpretation of the infimal convolution of functions. Let $f_1(x)$ and $f_2(x)$ be costs of producing x goods by production units U_1 and U_2 correspondingly. If we want to distribute optimally the production of a given x between U_1 and U_2, we have to solve the minimization problem

$$(f_1 \oplus f_2)(x) = \inf\{f_1(x_1) + f_2(x_2) \colon x_1 + x_2 = x\}.$$

THEOREM 3.14.– Let $f_i(x), i \in I$, be convex functions. Then the convex hull of their infimum

$$f(x) = (\text{conv}\,(\wedge_{i\in I} f_i))(x) = \text{conv}\,(\min_{i\in I} f_i(\cdot))(x)$$

$$= \inf\left\{\sum_{i\in I} \alpha_i f_i(x_i) : \alpha_i \geq 0, \sum_{i\in I} \alpha_i = 1, \sum_{i\in I} \alpha_i x_i = x\right\},$$

is a convex function.

Note that the epigraph of this function f is the convex hull of the union of epigraphs of functions $f_i(x), i \in I$.

THEOREM 3.15.– Let $f\colon \mathbb{R}^m \to \mathbb{R}$ be a convex function, and let $A : \mathbb{R}^m \to \mathbb{R}^n$ be a linear operator. Then the function $(Af)\colon \mathbb{R}^n \to \mathbb{R}$ defined by

$$(Af)(x) = \inf\{f(y)\colon y \in \mathbb{R}^m, Ay = x\}$$

(called the image of the function f under operator A) is convex.

The above theorems give us an effective tool for verification of the convexity of functions.

EXAMPLE 3.6.– Let $f_1, \dots, f_m$ be convex functions. Then the function

$$f(x) = \sum_{i=1}^{m} [\max\{0, f_i(x)\}]^q, \; q \geqslant 1$$

is convex. This function is used to solve linear programming problems by the penalty method.

Actually, the function $g_i(x) = \max\{0, f_i(x)\}$ is convex according to theorem 3.9. Since the function $\varphi(u) = u^q$, $q \geqslant 1$ is convex and non-decreasing on $\mathbb{R}_+$, then the function $\varphi(g_i(x))$ is convex according to theorem 3.11. So the function f is convex according to theorem 3.8.

EXAMPLE 3.7.– Let the numbers $\lambda_1 \geqslant 0, \dots, \lambda_n \geqslant 0$, $\sum_{i=1}^n \lambda_i = 1$. The Cobb-Douglas production function $f(x) = \prod_{i=1}^{n} x_i^{\lambda_i}$ is concave on $\mathbb{R}_+^n$.

Consider the set

$$Y = \left\{ y \in \mathbb{R}_+^n \,\middle|\, \prod_{i=1}^{n} y_i^{\lambda_i} = 1 \right\}$$

and the function $\varphi(x, y) = \sum_{i=1}^{n} \lambda_i x_i y_i$. For arbitrary $x > 0$, $y \in Y$, making use of the inequality from example 3.4, we have

$$f(x) = \prod_{i=1}^{n} x_i^{\lambda_i} = \prod_{i=1}^{n} (x_i y_i)^{\lambda_i} \leqslant \sum_{i=1}^{n} \lambda_i x_i y_i = \varphi(x, y).$$

At the same time, for the given $x > 0$ and $\vec{y} = f(x) \cdot (1/x_1, \dots, 1/x_n) \in Y$, we have $f(x) = \varphi(x, \vec{y})$. Therefore, $f(x) = \min_{y \in Y} \varphi(x, y)$ for all $x > 0$. Then, by theorem 3.9, the function f is concave on $\operatorname{int} \mathbb{R}_+^n$. Since f is continuous on $\mathbb{R}_+^n$, then the inequality that determines the concave functions (i.e. [3.1] with the opposite sign) holds true for all points from $\mathbb{R}^n_+$.

EXAMPLE 3.8.– *Piecewise linear functions.* Let $(a^i, b_i) \in \mathbb{R}^n \times \mathbb{R}$, $i = 1, \ldots, L$. The function

$$f(x) = \max_{i=1,\ldots,L} \{\langle a^i, x\rangle + b_i\}$$

is a piecewise linear function. This function is convex as the maximum of linear functions.

The inverse statement is also correct. Any piecewise linear convex function of not more than L segments of linearity can be represented in this form.

EXAMPLE 3.9.– *The sum of r largest components of a vector.* For $x \in \mathbb{R}^n$, denote by $x_{[i]}$ the ith largest component of the vector x, i.e. $x_{[1]}, x_{[2]}, \ldots, x_{[n]}$ are components of the vector x arranged in descending order. Then the function

$$f(x) = \sum_{i=1}^{r} x_{[i]},$$

i.e. the sum of r largest components of the vector x is convex. This can be seen by writing the function in the form

$$f(x) = \sum_{i=1}^{r} x_{[i]} = \max_{1 \leqslant i_1 < i_2 < \ldots < i_r \leqslant n} (x_{i_1} + \ldots + x_{i_r}),$$

where expression in the right side is the maximum among C_n^r of linear functions.

As a generalization of this example, we can show that the function

$$f(x) = \sum_{i=1}^{r} w_i x_{[i]}, \ w_1 \geqslant w_2 \geqslant \ldots \geqslant w_n \geqslant 0,$$

is convex.

EXAMPLE 3.10.– Minkowski function.

DEFINITION 3.11.– *Let X be a convex set and let $0 \in \operatorname{int} X$. Define the function*

$$\mu(x|X) = \inf\left\{\alpha \colon \alpha > 0, \frac{x}{\alpha} \in X\right\}.$$

This function $\mu(x|X)$ is defined for all $x \in \mathbb{R}^n$ and called the Minkowski function. It is also called the gauge function and is denoted by $\gamma_X(x)$. Since $0 \in \operatorname{int} X$, we have $\alpha^{-1}x \in X$ for large α for all x. Therefore, $\mu(x|X)$ is a finite number. In addition, by definition, $\mu(x|X) \geq 0$.

LEMMA 3.1.– Let X be a convex set and let $0 \in \text{int } X$. The Minkowski function has the following properties:

a) if $\lambda \geq 0$, then $\mu(\lambda x|X) = \lambda\mu(x|X)$;

b) for $x \in X$, we have $\mu(x|X) \leq 1$, and for $x \notin X$, we have $\mu(x|X) \geq 1$;

c) $\mu(x + y|X) \leq \mu(x|X) + \mu(y|X)$.

PROOF.– a) Let $\lambda\alpha_1 = \alpha$. We have

$$\mu(\lambda x|X) = \inf_{\alpha}\left\{\alpha : \alpha > 0, \frac{\lambda x}{\alpha} \in X\right\} = \inf_{\alpha_1}\left\{\lambda\alpha_1 : \alpha_1 > 0, \frac{x}{\alpha_1} \in X\right\}$$

$$= \lambda \inf_{\alpha_1}\left\{\alpha_1 : \alpha_1 > 0, \frac{x}{\alpha_1} \in X\right\} = \lambda\mu(x|X).$$

b) If $x \in X$, then $x/1 \in X$, so $\mu(x|X) \leq 1$. Let $x \notin X$. If $\mu(x|X) < 1$, then there exists $\alpha < 1$, such that $\alpha^{-1}x \in X$. Since $0 \in X$ and the set X is convex, then

$$x = (1 - \alpha) \cdot 0 + \alpha(\alpha^{-1}x) \in X.$$

This contradicts the assumption. So $\mu(x|X) \geq 1$ for $x \notin X$.

c) Let $\gamma > \mu(x|X) + \mu(y|X)$. We can find α and β such that $\gamma = \alpha + \beta, \alpha > \mu(x|X), \beta > \mu(y|X)$. From inequality $\alpha > \mu(x|X)$, it follows that $\alpha^{-1}x \in X$. Actually, by definition $\mu(x|X)$ there exists a number α_1 such that $\alpha > \alpha_1 > \mu(x|X)$ and $\alpha_1^{-1}x \in X$. Therefore, $\alpha^{-1}x = (1 - \alpha^{-1}\alpha_1) \cdot 0 + \alpha^{-1}\alpha_1(\alpha_1^{-1}x) \in X$ since the set X is convex and $0 \in X$. It is similarly proved that $\beta^{-1}y \in X$. Hence the validity of the relation

$$\frac{x + y}{\gamma} = \frac{x + y}{\alpha + \beta} = \frac{\alpha}{\alpha + \beta}(\alpha^{-1}x) + \frac{\beta}{\alpha + \beta}(\beta^{-1}y) \in X.$$

Therefore $\mu(x + y|X) \leq \gamma$.

But since γ is an arbitrary number greater that $\mu(x|X) + \mu(y|X)$, we have

$$\mu(x + y|X) \leq \mu(x|X) + \mu(y|X).$$

□

EXAMPLE 3.11.– *Indicator function* of a set. Let $C \subset \mathbb{R}^n$ be a set and let $C \neq \varnothing$. The indicator function of the set C is defined as follows:

$$\delta(x|C) = \begin{cases} 0, & x \in C, \\ +\infty, & x \notin C. \end{cases}$$

This function is (closed and) convex if and only if the set C is (closed and) convex. Indeed, the epigraph epi $\delta(\cdot|C) = C \times \mathbb{R}_+$ by definition.

EXAMPLE 3.12.– *Support function* of a set. Let $C \subset \mathbb{R}^n$ be a set and let $C \neq \varnothing$. The support function of the set C is defined as

$$\sigma(x|C) = \sup\{\langle x, y\rangle \mid y \in C\}$$

on the set $\left\{x \mid \sup_{y\in C}\langle x, y\rangle < \infty\right\}$.

For every $y \in C$, the function $\langle x, y\rangle$ is linear with respect to x. For this reason, the function $\sigma(x|C)$ is convex as the supremum of linear functions.

Since $\sigma(\alpha x|C) = \alpha\sigma(x|C)$ for $\alpha > 0$, the epigraph of a support function is not only closed and convex, but it is a cone in $\mathbb{R}^n \times \mathbb{R}$. Its domain is also a convex cone in $\mathbb{R}^n$.

EXAMPLE 3.13.– *Distance to the farthest point of the set.* Let $C \subset \mathbb{R}^n$. The distance (in any metric) to the farthest point of the set C

$$f(x) = \sup_{y\in C} \|x - y\|$$

is convex.

To prove this, we note that for a fixed y the function $\|x-y\|$ is convex with respect to the variable x. Since $f(x) = \sup_{y\in C} \|x-y\|$ is the supremum of a family of convex functions, it is convex with respect to the variable x.

EXAMPLE 3.14.– *The largest eigenvalue of a symmetric matrix.* The function $f(X) = \lambda_{\max}(X)$ on $\mathbb{S}^m$, where $\mathbb{S}^m$ is the set of symmetric matrices of dimension $m \times m$, is convex.

To see this, we write the function f in the form

$$f(X) = \sup\{\langle Xy, y\rangle : \|y\| = 1\},$$

i.e. in the form of the supremum of linear functions of X.

EXAMPLE 3.15.– *Norm of matrix.* Consider the function $f(X) = \|X\|_2$ on the set of matrices of dimension $p \times q$, where $\|\cdot\|_2$ denotes the spectral norm or the largest eigenvalue. The convexity of f follows from the identity

$$f(X) = \sup\{\langle Xv, u\rangle : \|u\| \leqslant 1, \|v\| \leqslant 1\}.$$

The expression in the right side is the supremum of the linear functions of X.

EXAMPLE 3.16.– *Induced norm of matrix.* Let $\|\cdot\|_p$ and $\|\cdot\|_q$ be norms on $\mathbb{R}^p$ and $\mathbb{R}^q$ correspondingly. Then the induced norm

$$\|X\|_{p,q} = \sup_{v\neq 0} \frac{\|Xv\|_p}{\|v\|_q}$$

is convex on the set of matrices of dimension $p \times q$, since it can be represented in the form

$$f(X) = \sup\{\langle u, Xv\rangle : \|u\|_{p^*} \leqslant 1, \|v\|_q \leqslant 1\},$$

where $\|\cdot\|_{p^*}$ is a conjugate to $\|\cdot\|_p$ norm.

EXAMPLE 3.17.– Let A be a symmetric matrix of dimension $n\times n$. Then the quadratic form

$$f(x) = \langle Ax, x\rangle$$

is a convex function if and only if A is positive semidefinite, i.e. its eigenvalues are all non-negative. Let $\lambda_1 \geq \cdots \geq \lambda_n \geq 0$ be these eigenvalues. A basis can be formed with the corresponding eigenvector and, as result, we have

$$\lambda_n\|x\|^2 \leq \langle Ax, x\rangle \leq \lambda_1\|x\|^2.$$

From the first inequality, we have that

$$f(\alpha x^1 + (1-\alpha)x^2) \leq \alpha f(x^1) + (1-\alpha)f(x^2) - \lambda_n\alpha(1-\alpha)\|x^1 - x^2\|^2.$$

That is, f is strongly convex with modulus of convexity $\lambda_n > 0$ if A is positive definite, while f is not even strictly convex when A is degenerate.

EXAMPLE 3.18.– Let A_1 and A_2 be symmetric positive definite matrices of dimension $n \times n$. Consider two quadratic forms

$$f_1(x) = \langle A_1x, x\rangle, \quad f_2(x) = \langle A_2x, x\rangle.$$

Their infimal convolution is of the form

$$(f_1 \oplus f_2)(x) = \inf_y \{\langle A_1y, y\rangle + \langle A_2(x-y), (x-y)\rangle\}.$$

The minimum can be calculated explicitly to give

$$(f_1 \oplus f_2)(x) = \langle A_{12}x, x\rangle, \quad A_{12} = \left(A_1^{-1} + A_2^{-1}\right)^{-1}.$$

This example of two convex quadratic functions can be extended to a sequence of quadratic functions.

3.3. Criteria of convexity of differentiable functions

In this section, we describe the necessary and sufficient conditions of convexity and strong convexity of differentiable functions $f\colon \mathbb{R}^n \to \mathbb{R}$.

THEOREM 3.16.– Let $f\colon \mathbb{R}^n \to \mathbb{R}$ be a differentiable on an open set $U \subset \mathbb{R}^n$ function and let $X \subset U$ be a convex set. The following statements hold true:

i) function f is convex on X if and only if

$$f(x) - f(\hat{x}) \geqslant \langle f'(\hat{x}), x - \hat{x} \rangle \quad \text{for all} \quad x, \hat{x} \in X; \qquad [3.6]$$

ii) function f is strictly convex on X if and only if

$$f(x) - f(\hat{x}) > \langle f'(\hat{x}), x - \hat{x} \rangle \text{ for all } x, \hat{x} \in X,\ x \neq \hat{x}; \qquad [3.7]$$

iii) function f is strongly convex on X with the module of convexity $\theta > 0$ if and only if

$$f(x) - f(\hat{x}) \geqslant \langle f'(\hat{x}), x - \hat{x} \rangle + \theta \left\| x - \hat{x} \right\|^2 \text{ for all } x, \hat{x} \in X. \qquad [3.8]$$

PROOF.– Let f be a strongly convex function with the module of convexity $\theta \geqslant 0$. By definition, for all $x, \hat{x} \in X$, $\lambda \in [0, 1]$, the following inequality holds:

$$f\left(\lambda x + (1 - \lambda)\hat{x}\right) \leqslant \lambda f(x) + (1 - \lambda) f(\hat{x}) - \theta \lambda (1 - \lambda) \left\| x - \hat{x} \right\|^2.$$

Making use of the differentiability of the function f at point $\hat{x}$, we have

$$f(x) - f(\hat{x}) - \theta (1 - \lambda) \left\| x - \hat{x} \right\|^2 \geqslant \frac{f\left(\hat{x} + \lambda (x - \hat{x})\right) - f(\hat{x})}{\lambda}$$

$$= \frac{\langle f'(\hat{x}), \lambda (x - \hat{x}) \rangle + o(\lambda)}{\lambda} = \langle f'(\hat{x}), x - \hat{x} \rangle + \frac{o(\lambda)}{\lambda}.$$

The passage to the limit as $\lambda \to 0$ in this inequality leads to inequality [3.8].

Let inequality [3.8] hold true. For arbitrary x^1, $x^2 \in X$, $\lambda \in [0, 1]$ set $\hat{x} = \lambda x^1 + (1 - \lambda) x^2 \in X$. It follows from the inequality [3.8] that

$$f(x^1) - f(\hat{x}) \geqslant \langle f'(\hat{x}), x^1 - \hat{x} \rangle + \theta \|x^1 - \hat{x}\|^2,$$

$$f(x^2) - f(\hat{x}) \geqslant \langle f'(\hat{x}), x^2 - \hat{x} \rangle + \theta \|x^2 - \hat{x}\|^2.$$

Note that

$$\|x^1 - \hat{x}\| = (1 - \lambda)\|x^1 - x^2\|, \quad \|x^2 - \hat{x}\| = \lambda \|x^1 - x^2\|$$

We multiply the first of the prescribed inequalities by λ, the second inequality we multiply by $(1 - \lambda)$ and add them. We get

$$\lambda f(x^1) + (1 - \lambda) f(x^2) - f(\hat{x}) - \theta \lambda (1 - \lambda) \|x^1 - x^2\|^2$$

$$\geqslant \langle f'(\hat{x}), \lambda x^1 + (1 - \lambda) x^2 - \hat{x} \rangle = \langle f'(\hat{x}), 0 \rangle = 0.$$

This means that inequality [3.2], according to the definition of a strong convex function, holds true. The proofs of convexity and strict convexity are similar. □

As a corollary, we have the following statement.

THEOREM 3.17.– Let f be a convex function on an open convex set $X \subset \mathbb{R}^n$. Let the function f be differentiable at point $\hat{x} \in X$. Then

$$f(x) - f(\hat{x}) \geqslant \langle f'(\hat{x}), x - \hat{x} \rangle \quad \text{for all} \quad x \in X. \qquad [3.9]$$

Note that the graph of a linear function $l(x) = f(\hat{x}) + \langle f'(\hat{x}), x - \hat{x} \rangle$ is called a tangential hyperplane to the graph of a function f at point $(\hat{x}, f(\hat{x}))$. Relation [3.9] means that the graph of the function f is disposed above the tangential hyperplane at the point $(\hat{x}, f(\hat{x}))$.

Based on the proved theorem, we can obtain the following criterion of convexity of a function in terms of the first derivatives.

THEOREM 3.18.– Let $f\colon \mathbb{R}^n \to \mathbb{R}$ be a differentiable on an open set $U \subset \mathbb{R}^n$ function and let $X \subset U$ be a convex set. The following statements hold true:

i) the function f is convex on X if and only if

$$\langle f'(x) - f'(\hat{x}), x - \hat{x} \rangle \geqslant 0 \quad \text{for all} \quad x, \hat{x} \in X; \qquad [3.10]$$

ii) the function f is strictly convex on X if and only if

$$\langle f'(x) - f'(\hat{x}), x - \hat{x} \rangle > 0 \quad \text{for all} \quad x, \hat{x} \in X,\ x \neq \hat{x}; \qquad [3.11]$$

iii) the function f is strongly convex on X with the module of convexity $\theta > 0$ if and only if

$$\langle f'(x) - f'(\hat{x}), x - \hat{x} \rangle \geqslant 2\theta \|x - \hat{x}\|^2 \text{ for all} \quad x,\ \hat{x} \in X. \qquad [3.12]$$

PROOF.– Let f be strongly convex on X with the module of convexity $\theta \geqslant 0$. It follows from theorem 3.16 (inequality [3.8]) that

$$f(x) - f(\hat{x}) \geqslant \langle f'(\hat{x}), x - \hat{x} \rangle + \theta \|x - \hat{x}\|^2,$$

$$f(\hat{x}) - f(x) \geqslant \langle f'(x), \hat{x} - x \rangle + \theta \|x - \hat{x}\|^2$$

for all $x,\ \hat{x} \in X$. Adding these inequalities, we obtain inequality [3.12].

Let, on the contrary, inequality [3.12] hold true. Then

$$f(x) - f(\hat{x}) - \langle f'(\hat{x}), x - \hat{x} \rangle$$

$$= \int_0^1 \langle f'(\hat{x} + \alpha(x - \hat{x})), x - \hat{x} \rangle d\alpha - \langle f'(\hat{x}), x - \hat{x} \rangle$$

$$= \int_0^1 \frac{1}{\alpha} \langle f'(\hat{x} + \alpha(x - \hat{x})) - f'(\hat{x}), \alpha(x - \hat{x}) \rangle d\alpha$$

$$\geqslant \int_0^1 \frac{1}{\alpha} 2\theta \|\alpha(x - \hat{x})\|^2 d\alpha = \theta \|x - \hat{x}\|^2,$$

where the Newton–Leibniz formula is used in the first equation, and formula [3.12] for points $\hat{x}$ and $x^\alpha = \hat{x} + \alpha(x - \hat{x}) \in X$, $0 < \alpha \leqslant 1$, is used in inequality. So [3.8] holds true. It means that the function f is strongly convex with the module of convexity $\theta \geqslant 0$. □

For the function f of a real argument ($n = 1$), condition [3.12] has the form $f'(x) - f'(\hat{x}) \geqslant 2\theta(x - \hat{x})$ for all $x, \hat{x} \in X$, $x \geqslant \hat{x}$. In the case of $\theta = 0$ (case of a convex function), it means that the derivative $f'(x)$ is monotonically non-decreasing on X.

Next, we describe criteria of convexity in terms of the second derivatives.

THEOREM 3.19.– Let $f\colon \mathbb{R}^n \to \mathbb{R}$ be two times continuously differentiable on a convex set $X \subset \mathbb{R}^n$ function, and let $\operatorname{int} X \neq \varnothing$. The following statements hold true:

i) the function f is convex on X if and only if

$$\langle f''(\hat{x})h, h \rangle \geqslant 0 \quad \text{forall} \quad \hat{x} \in X,\ h \in \mathbb{R}^n; \qquad [3.13]$$

ii) the function f is strictly convex on X if

$$\langle f''(\hat{x})h, h \rangle > 0 \quad \text{forall} \quad \hat{x} \in X,\ h \in \mathbb{R}^n, h \neq 0; \qquad [3.14]$$

iii) the function f is strongly convex with the module of convexity $\theta > 0$ on X if and only if

$$\langle f''(\hat{x})h, h \rangle \geqslant 2\theta \|h\|^2 \quad \text{forall} \quad \hat{x} \in X,\ h \in \mathbb{R}^n. \qquad [3.15]$$

PROOF.– Let a function f be strongly convex with the module of convexity $\theta \geqslant 0$. We first suppose that $\hat{x} \in \operatorname{int} X$. In this case for all $h \in \mathbb{R}^n$, we have $\hat{x} + \alpha h \in X$ for small enough $\alpha > 0$. Making use of the differentiability of the function f at point $\hat{x}$, we can write

$$f(\hat{x} + \alpha h) = f(\hat{x}) + \langle f'(\hat{x}), \alpha h \rangle + \frac{1}{2} \langle f''(\hat{x}) \alpha h, \alpha h \rangle + o(\alpha^2).$$

It follows from this relation that

$$\frac{\alpha^2}{2}\langle f''(\hat{x})h, h\rangle + o(\alpha^2) = f(\hat{x}+\alpha h) - f(\hat{x}) - \langle f'(\hat{x}), \alpha h\rangle \geqslant \theta\alpha^2\|h\|^2,$$

where inequality is a consequence of inequality [3.8]. Therefore

$$\frac{1}{2}\langle f''(\hat{x})h, h\rangle + \frac{o(\alpha^2)}{\alpha^2} \geqslant \theta\|h\|^2.$$

Transition to the limit as $\alpha \to 0$ in this inequality leads to inequality [3.15].

Consider now the general case $\hat{x} \in X$. Since $X \subset \overline{X} = \overline{\operatorname{int} X}$, we can find a sequence of points $x^k \in \operatorname{int} X$, $k = 1, 2, \ldots$, which converges to $\hat{x}$. For these points, we already proved that for arbitrary $h \in \mathbb{R}^n$, we have $\langle f''(x^k)h, h\rangle \geqslant 2\theta\|h\|^2$, $k = 1, 2, \ldots$. It follows from the continuity of $f''(x)$ at point $\hat{x}$ that the sequence of matrices $f''(x^k)$, $k = 1, 2, \ldots$ converges to $f''(\hat{x})$. Making use of these properties we get inequality [3.15].

Let, on the contrary, equality [3.15] hold true. For arbitrary $x, \hat{x} \in X$ take $h = x - \hat{x}$. Then, making use of Taylor's theorem, with a residual term in the Lagrange form and formula [3.15], for some $\bar{\alpha} \in (0, 1)$ we have

$$f(\hat{x}+h) - f(\hat{x}) - \langle f'(\hat{x}), h\rangle = \frac{1}{2}\langle f''(\hat{x}+\bar{\alpha}h)h, h\rangle \geqslant \theta\|h\|^2.$$

This means that relation [3.8] holds true. Consequently, the function f is strongly convex on X. □

REMARK 3.1.– The sufficient condition in (ii) is not the necessary condition. For example, the function $f(x) = x^4$ is strictly convex on $\mathbb{R}$, but $f''(0) = 0$.

COROLLARY 3.2.– Let A be a symmetric matrix of $n \times n$ dimension, and let $b \in \mathbb{R}^n$. The quadratic function $f(x) = \langle Ax, x\rangle + \langle b, x\rangle$ is convex (strongly convex) on $\mathbb{R}^n$ if and only if the matrix A is non-negative (positive) definite.

PROOF.– The validity of the statement follows from the proved theorem since $f''(x) = 2A$. Note that in the case where the matrix A is positive definite, the number $m = \min\limits_{\|h\|=1} \langle Ah, h\rangle$ is positive. Inequality [3.15] holds true with $\theta = m$. □

For a function f of one real argument ($n = 1$) condition [3.15] has the form

$$f''(\hat{x}) \geqslant 2\theta \quad \text{for all} \quad \hat{x} \in X. \qquad [3.16]$$

For $\theta = 0$, this means that the second derivative $f''(x)$ is non-negative on X. For $\theta > 0$, it is necessary that $f''(x)$ is no less than a positive constant. It follows from this condition that the function $f(x) = x^4$ is not strongly convex on $\mathbb{R}$, since $f''(0) = 0$. But on the set $X = [\alpha, +\infty]$ with $\alpha > 0$, this function $f(x) = x^4$ is strongly convex. The function $f(x) = e^x$ is not strongly convex on $\mathbb{R}$ also, since $f''(x) = e^x \to 0$ as $x \to -\infty$. But on the set $X^1 = [\alpha, +\infty]$, where α is an arbitrary number, it is strongly convex. The last example shows that the positive definiteness of the matrix $f''(x)$ on X does not follow the strong convexity of the function f.

The proved theorems in combination with the Sylvester criterion provide convenient devices for checking the convexity of functions of several variables.

EXAMPLE 3.19.– Consider the quadratic function of two variables $f(x) = ax_1^2 + bx_1x_2 + cx_2^2$. The matrix of the second derivatives is of the form

$$f''(x) = \begin{pmatrix} 2a & b \\ b & 2c \end{pmatrix}.$$

According to the Sylvester criterion, this matrix is non-negative definite if $a \geqslant 0$, $c \geqslant 0$, $4ac \geqslant b^2$, and it is positive definite if $a > 0$, $4ac > b^2$. Therefore in the first case, the function f is convex and in the second case it is strongly convex on $\mathbb{R}^2$.

EXAMPLE 3.20.– Consider the function of two variables $f(x_1, x_2) = x_1^2/x_2$ on the set $X = \{x = (x_1, x_2) \in \mathbb{R}^2 \,|\, x_2 > 0\}$. The matrix of the second derivative is of the form

$$f''(x) = \begin{pmatrix} 2/x_2 & -2x_1/x_2^2 \\ -2x_1/x_2^2 & 2x_1^2/x_2^3 \end{pmatrix}.$$

According to the Sylvester criterion, this matrix is non-negative definite for all $x \in X$. Therefore, the function f is convex on X. Note that proof of the convexity of this function based on definition 3.1 would lead to rather complicated calculations.

EXAMPLE 3.21.– *Logarithm of the sum of exponents.* The function

$$f(x) = \ln\left(e^{x_1} + \ldots + e^{x_n}\right)$$

is convex on $\mathbb{R}^n$. This function is a smooth approximation of the maximum $\max\{x_1, \ldots, x_n\}$, since

$$\max\{x_1, \ldots, x_n\} \leqslant f(x) \leqslant \max\{x_1, \ldots, x_n\} + \ln n$$

for all x. The matrix of the second derivatives is of the form

$$f''(x) = \frac{1}{\left(\sum_{i=1}^n z_i\right)^2} \left(\left(\sum_{i=1}^n z_i \right) \mathrm{diag}(\vec{z}) - \vec{z} \cdot \vec{z}^{\top} \right),$$

where $\vec{z} = (e^{x_1}, \ldots, e^{x_n})$. To check the non-negative definiteness of $f''(x)$, we must show that $\langle f''(x)v, v\rangle \geqslant 0$ for all v, that is

$$\langle f''(x)v, v\rangle = \frac{1}{(\sum_{i=1}^{n} z_i)^2}\left(\left(\sum_{i=1}^{n} z_i\right)\left(\sum_{i=1}^{n} v_i^2 z_i\right) - \left(\sum_{i=1}^{n} v_i z_i\right)^2\right) \geqslant 0.$$

But this inequality follows from the Cauchy–Bunyakovsky inequality $\langle a, a\rangle\langle b, b\rangle \geqslant \langle a, b\rangle^2$ for vectors with components $a_i = v_i\sqrt{z_i}$, $b_i = \sqrt{z_i}$.

EXAMPLE 3.22.– *Geometric mean.* The function (geometric mean)

$$f(x) = \left(\prod_{i=1}^{n} x_i\right)^{1/n}$$

is concave on $\operatorname{int} \mathbb{R}^n_+$.

Consider the matrix $f''(x) = \left(f''_{jk}(x)\right)_{j,\,k=1}^{n}$, where

$$f''_{kk}(x) = -(n-1)\frac{\left(\prod_{i=1}^{n} x_i\right)^{1/n}}{n^2 x_k^2}, \quad f''_{jk}(x) = \frac{\left(\prod_{i=1}^{n} x_i\right)^{1/n}}{n^2 x_j x_k} \quad (k \neq j).$$

This matrix can be represented in the form

$$f''(x) = -\frac{\prod_{i=1}^{n} x_i^{1/n}}{n^2}\left(n \operatorname{diag}\left(\frac{1}{x_1^2}, \ldots, \frac{1}{x_n^2}\right) - \vec{q}\cdot\vec{q}^{\mathsf{T}}\right),$$

where $\vec{q} = (q_1, \ldots, q_n)$, $q_i = 1/x_i$. We have to show that $f''(x) \leqslant 0$, namely

$$\langle f''(x)v, v\rangle = -\frac{\prod_{i=1}^{n} x_i^{1/n}}{n^2}\left(n\sum_{i=1}^{n}\frac{v_i^2}{x_i^2} - \left(\sum_{i=1}^{n}\frac{v_i}{x_i}\right)^2\right) \leqslant 0$$

for all v.

This inequality follows from the Cauchy–Bunyakovsky inequality $\langle a, a\rangle\langle b, b\rangle \geqslant \langle a, b\rangle^2$ for vectors with components $a_i = 1$, $b_i = v_i/x_i$.

EXAMPLE 3.23.– *Logarithm of determinant.* The function

$$f(X) = \ln\det X^{-1}$$

is convex in $\mathbb{S}^n_{++}$.

We can check the convexity of the function $f(X) = \ln\det X^{-1}$ by considering it on a straight line $X = Z + tV$, where $Z, V \in \mathbb{S}^n$. Define the function

$g(t) = f(Z + tV)$. We will consider it for such values of t that $Z + tV > 0$. We can assume that $t = 0$ belongs to this set, so Z is positive definite. We have

$$g(t) = -\ln\det(Z + tV) = -\ln\det Z^{1/2}\left(I + tZ^{-1/2}VZ^{1/2}\right)Z^{1/2}$$

$$= -\sum_{i=1}^{n}\ln(1 + t\lambda_i) - \ln\det Z,$$

where $\lambda_1, \ldots, \lambda_n$ are eigenvalues of the matrix $Z^{-1/2}VZ^{1/2}$. Therefore

$$g'(t) = -\sum_{i=1}^{n}\frac{\lambda_i}{1 + t\lambda_i}, \quad g''(t) = \sum_{i=1}^{n}\frac{\lambda_i^2}{(1 + t\lambda_i)^2}.$$

Since $g''(t) \geqslant 0$, we conclude that the function f is convex.

3.4. Continuity and differentiability of convex functions

Inequality [3.1], which defines a convex function, turns out to be so strong that it ensures continuity and differentiability of a convex function in all directions at each interior point of the set of definition.

THEOREM 3.20.– Let $f : \mathbb{R}^n \to \mathbb{R}^1$ be a convex function on a convex set $X \subset \mathbb{R}^n$. The function $f(x)$ is continuous at each point $\hat{x} \in \operatorname{ri} X$.

PROOF.– First, suppose that (1) $\hat{x} = 0$, $f(\hat{x}) = 0$; (2) $\operatorname{int} X \neq \varnothing$. Since $0 \in \operatorname{ri} X = \operatorname{int} X$, then there is such a small number $r > 0$, that the hypercube $K = \{x \in \mathbb{R}^n \mid -r \leqslant x_j \leqslant r,\ j = 1, \ldots, n\}$, which is a neighbourhood of zero, belongs to X. Let $x^1, \ldots, x^m$, where $m = 2^n$ indicates its vertices, that is, points of the form $x^i = (\pm r, \ldots, \pm r)$. Denote $\alpha = \max_{i=1,\ldots,m}\{f(x^i)\}$. Each point $x \in K$ can be represented as a convex combination of points $x^1, \ldots, x^m$, that is $x = \sum_{i=1}^m \lambda_i x^i$, where $\lambda_1 \geqslant 0, \ldots, \lambda_m \geqslant 0$, $\sum_{i=1}^m \lambda_i = 1$ (we can write explicit expressions for λ_i). Then, making use of Jensen's inequality [3.3], we have

$$f(x) \leqslant \sum_{i=1}^{m}\lambda_i f(x^i) \leqslant \alpha\sum_{i=1}^{m}\lambda_i = \alpha. \qquad [3.17]$$

Therefore, the function f is bounded from above on K.

Consider a number $\varepsilon \in (0, 1]$ and the neighborhood of zero $U_\varepsilon = \varepsilon K$. For each point $x \in U_\varepsilon$, making use of convexity of the function f, condition $f(0) = 0$ and estimate [3.17] for points $\pm x/\varepsilon \in K$, we have the inequalities

$$f(x) = f\left(\varepsilon\frac{x}{\varepsilon} + (1 - \varepsilon)0\right) \leqslant \varepsilon f\left(\frac{x}{\varepsilon}\right) + (1 - \varepsilon)f(0) \leqslant \varepsilon\alpha,$$

$$0 = f(0) = f\left(\frac{1}{1+\varepsilon}\,x + \frac{\varepsilon}{1+\varepsilon}\left(-\frac{x}{\varepsilon}\right)\right) \leqslant \frac{1}{1+\varepsilon}f(x) + \frac{\varepsilon}{1+\varepsilon}\,\alpha.$$

It follows from these inequalities that $|f(x)| \leqslant \varepsilon\alpha$. Therefore, the function f is continuous at point $\hat{x} = 0$.

Let now $\operatorname{int} X = \varnothing$, but still $\hat{x} = 0$, $f(0) = 0$. Consider a linear homeomorphic mapping $F : \mathbb{R}^m \to \operatorname{Lin} X$. Put $\Lambda = F^{-1}(X)$. Since $0 \in \operatorname{ri} X$ is an interior point of the set X in $\operatorname{Lin} X = \operatorname{aff} X$, then $0 = F^{-1}(0)$ is an interior point of the set Λ in $\mathbb{R}^m$, that is $0 \in \operatorname{int} \Lambda$. Define the function $\varphi(\lambda) = f(F(\lambda))$ on Λ. Due to the linearity F and F^{-1}, we have that Λ is a convex set, and φ is a convex function on Λ. We have $\varphi(0) = 0$, since $f(0) = 0$. Then, as proved already, φ is continuous at zero. Since $f(x) = \varphi(F^{-1}(x))$, the function f is continuous at point $\hat{x} = 0$ as superposition of continuous functions. Therefore, the theorem is proved in the case where $\hat{x} = 0$, $f(0) = 0$. It remains to note that the general case is reduced to this case with the help of function $\Phi(y) = f(y + \hat{x}) - f(\hat{x})$ on the set $Y = X - \hat{x}$. □

REMARK 3.2.– For relative boundary points, this theorem is not true. For example, the function

$$f(x) = \begin{cases} x, \text{ if } & x > 0, \\ 1, \text{ if } & x = 0 \end{cases}$$

is convex on $\mathbb{R}_+$, but it is discontinuous at zero point.

Before proving the differentiability of a convex function with respect to all directions, we note the following.

Let $\hat{x} \in \operatorname{ri} X$ and $h \in \operatorname{Lin} X$. Then $U_\varepsilon(\hat{x}) \cap \operatorname{aff} X \subset X$ for some $\varepsilon > 0$ and $\hat{x} + \alpha h \in \operatorname{aff} X$ for all α. Hence $\hat{x} + \alpha h \in U_\varepsilon(\hat{x}) \cap \operatorname{aff} X \subset X$ for small enough α. That is why the sets

$$A = \{\alpha > 0 \,|\, \hat{x} + \alpha h \in X\}, \quad B = \{\alpha < 0 \,|\, \hat{x} + \alpha h \in X\} \qquad [3.18]$$

are non-empty.

THEOREM 3.21.– Let f be a convex function on a convex set $X \subset \mathbb{R}^n$, and let $\hat{x} \in \operatorname{ri} X$ and $h \in \operatorname{Lin} X$. The following statements hold true:

1) the function

$$\psi(\alpha) = \frac{f(\hat{x} + \alpha h) - f(\hat{x})}{\alpha}$$

is monotonically non-decreasing and bounded from below on the set A;

2) the quantity $f'(\hat{x}; h) = \lim\limits_{\alpha \to 0+} \psi(\alpha)$ exists, is finite and, besides,

$$f'(\hat{x}; h) \leqslant \psi(\alpha) \quad \text{for all} \quad \alpha \in A. \qquad [3.19]$$

PROOF.– Note that (2) follows from (1) by the theorem about the limit of a monotone function.

To prove statement (1), consider arbitrary α, $\alpha' \in A$, $\alpha \leqslant \alpha'$. It follows from the convexity of the function f that

$$f(\hat{x} + \alpha h) = f\left(\left(1 - \frac{\alpha}{\alpha'}\right)\hat{x} + \frac{\alpha}{\alpha'}(\hat{x} + \alpha' h)\right)$$
$$\leqslant \left(1 - \frac{\alpha}{\alpha'}\right) f(\hat{x}) + \frac{\alpha}{\alpha'} f(\hat{x} + \alpha' h).$$

Therefore $\psi(\alpha) \leqslant \psi(\alpha')$, that is $\psi(\alpha)$ is monotonically non-decreasing on A.

In the same manner for arbitrary $\alpha \in A$, $\alpha' \in B$, we have

$$f(\hat{x}) = f\left(\frac{\alpha}{\alpha - \alpha'}(\hat{x} + \alpha' h) + \frac{-\alpha'}{\alpha - \alpha'}(\hat{x} + \alpha h)\right)$$
$$\leqslant \frac{\alpha}{\alpha - \alpha'} f(\hat{x} + \alpha' h) + \frac{-\alpha'}{\alpha - \alpha'} f(\hat{x} + \alpha h).$$

Therefore $\psi(\alpha) \geqslant \psi(\alpha')$. This means that the function $\psi(\alpha)$ is bounded from below on A. □

REMARK 3.3.– From the proof of this theorem, it follows that for a point $\hat{x} \in X \setminus \operatorname{ri} X$ and vectors h such that the set A from [3.18] is non-empty, the function $f'(\hat{x}; h)$ also exists, but it can be equal to $-\infty$. This situation is realized, for example, for function $f(x) = -\sqrt{1 - x^2}$, which is convex on $[-1, 1]$, point $\hat{x} = 1$ and vector $h = -1$.

LEMMA 3.2.– Inequalities for convex functions. Let $g(\alpha)$, $\alpha \in \mathbb{R}$, be a convex function, and let $\alpha_0 < \alpha_1 < \alpha_2$; $\alpha_0, \alpha_1, \alpha_2 \in \operatorname{dom} g$. Then the following inequalities hold true:

$$\frac{g(\alpha_2) - g(\alpha_0)}{\alpha_2 - \alpha_0} \geq \frac{g(\alpha_1) - g(\alpha_0)}{\alpha_1 - \alpha_0},$$

$$\frac{g(\alpha_1) - g(\alpha_0)}{\alpha_1 - \alpha_0} \leq \frac{g(\alpha_2) - g(\alpha_1)}{\alpha_2 - \alpha_1}.$$

PROOF.– Take

$$\lambda_1 = \frac{\alpha_1 - \alpha_0}{\alpha_2 - \alpha_0}, \quad \lambda_2 = 1 - \lambda_1 = \frac{\alpha_2 - \alpha_1}{\alpha_2 - \alpha_0}.$$

Then

$$\lambda_1 \alpha_2 + \lambda_2 \alpha_0 = \frac{\alpha_1 - \alpha_0}{\alpha_2 - \alpha_0}\alpha_2 + \frac{\alpha_2 - \alpha_1}{\alpha_2 - \alpha_0}\alpha_0 = \alpha_1.$$

Therefore

$$g(\alpha_1) = g(\lambda_1\alpha_2 + \lambda_2\alpha_0) \leq \frac{\alpha_1 - \alpha_0}{\alpha_2 - \alpha_0} g(\alpha_2) + \frac{\alpha_2 - \alpha_1}{\alpha_2 - \alpha_0} g(\alpha_0).$$

There are several ways to transform the resulting inequality:

1) Subtracting from both parts $g(\alpha_0)$, we obtain

$$g(\alpha_1) - g(\alpha_0) \leq \frac{\alpha_1 - \alpha_0}{\alpha_2 - \alpha_0} (g(\alpha_2) - g(\alpha_0).$$

Then, after dividing by $\alpha_1 - \alpha_0$, the first inequality of the lemma follows.

2) Since $\lambda_1 + \lambda_2 = 1$, then

$$\frac{\alpha_1 - \alpha_0}{\alpha_2 - \alpha_0} g(\alpha_1) + \frac{\alpha_2 - \alpha_1}{\alpha_2 - \alpha_0} g(\alpha_1) \leq \frac{\alpha_1 - \alpha_0}{\alpha_2 - \alpha_0} g(\alpha_2) + \frac{\alpha_2 - \alpha_1}{\alpha_2 - \alpha_0} g(\alpha_0),$$

or

$$\frac{\alpha_2 - \alpha_1}{\alpha_2 - \alpha_0} [g(\alpha_1) - g(\alpha_0)] \leq \frac{\alpha_1 - \alpha_0}{\alpha_2 - \alpha_0} [g(\alpha_2) - g(\alpha_1)].$$

From this, the second inequality of the lemma follows. □

REMARK 3.4.– Note that the first of the inequalities proved in the lemma shows that the function

$$\frac{g(\alpha) - g(\alpha_0)}{\alpha - \alpha_0}$$

is monotonically non-decreasing with increasing $\alpha \geq \alpha_0$.

THEOREM 3.22.– (*Theorem on separating linear function*) Let X_1 and X_2 be convex sets in $\mathbb{R}^n$, let $f_1(x)$ be a convex function on X_1, let $f_2(x)$ be a concave function on X_2, let $\operatorname{ri} X_1 \cap \operatorname{ri} X_2 \neq \varnothing$ and let

$$f_1(x) \geqslant f_2(x) \quad \text{for all} \quad x \in X_1 \cap X_2. \qquad [3.20]$$

Then there exists a (separating) linear function $l(x) = \langle a, x \rangle + b$, such that

$$f_1(x) \geqslant l(x) \quad \text{for all} \quad x \in X_1, \qquad [3.21]$$

$$l(x) \geqslant f_2(x) \quad \text{for all} \quad x \in X_2. \qquad [3.22]$$

Geometrically this means that the hyperplane, which is a graph of the functions $l(x) = \langle a, x \rangle + b$, passes under the graph of the function $f_1(x)$ and over the graph of the function $f_2(x)$ (Figure 3.4).

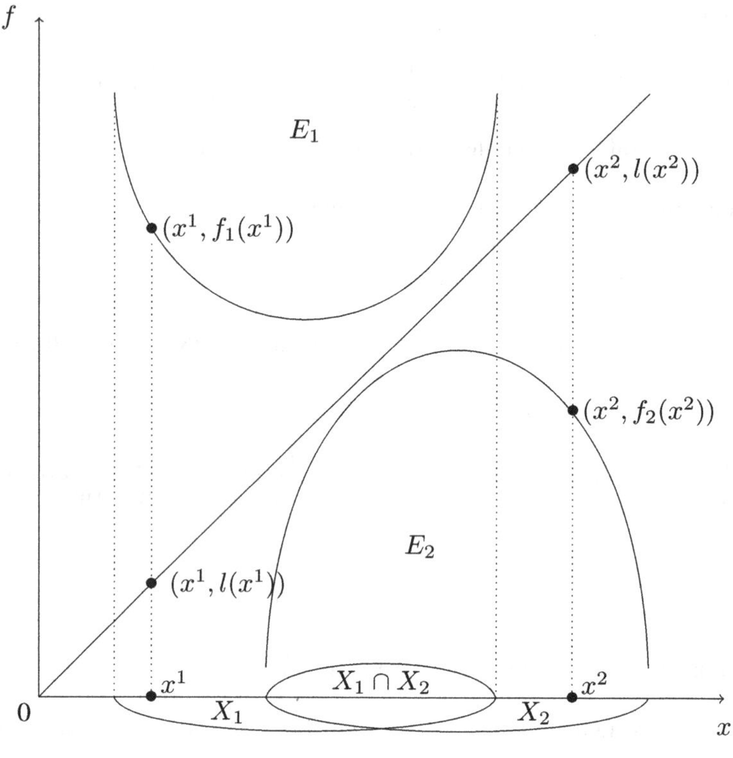

Figure 3.4. *Separating linear function*

PROOF.– Consider the following sets:

$$E_1 = \{(x, \beta) \in \mathbb{R}^n \times \mathbb{R} \,|\, x \in X_1,\ f_1(x) < \beta\},$$

$$E_2 = \{(x, \beta) \in \mathbb{R}^n \times \mathbb{R} \,|\, x \in X_2,\ \beta < f_2(x)\}.$$

It follows from [3.20] that $E_1 \cap E_2 = \varnothing$. Then $\operatorname{ri} E_1 \cap \operatorname{ri} E_2 = \varnothing$. First we show that the set E_1 is convex. For arbitrary $(x^1, \beta^1) \in E_1$, $(x^2, \beta^2) \in E_1$ and $\lambda \in [0, 1]$ take $\bar{x} = \lambda x^1 + (1 - \lambda)x^2$, $\bar{\beta} = \lambda\beta^1 + (1 - \lambda)\beta^2$. It follows from the convexity of the set X_1 and the function $f_1(x)$ that $\bar{x} \in X_1$, $f_1(\bar{x}) \leqslant \lambda f_1(x^1) + (1 - \lambda) f_1(x^2) < \bar{\beta}$, that is $(\bar{x}, \bar{\beta}) = \lambda(x^1, \beta^1) + (1 - \lambda)(x^2, \beta^2) \in E_1$. Therefore, the set E_1 is convex. Analogically, making use of the convexity of the set X_2 and concavity of the function $f_2(x)$, we prove that the set E_2 is convex. Then, following Fenchel's separation theorem, the sets E_1 and E_2 are properly separated, that is, there is a vector $(p, \lambda) \in \mathbb{R}^n \times \mathbb{R}$ and a number α such that

$$\langle p, x^1 \rangle + \lambda\beta^1 \geqslant \alpha \geqslant \langle p, x^2 \rangle + \lambda\beta^2 \qquad [3.23]$$

for all $x^1 \in X_1$, $\beta^1 > f_1(x^1)$, $x^2 \in X_2$, $\beta^2 < f_2(x^2)$ and, besides,

$$\langle p, \bar{x}^1 \rangle + \lambda\bar{\beta}^1 > \langle p, \bar{x}^2 \rangle + \lambda\bar{\beta}^2$$

for some $\bar{x}^1 \in X_1$, $\bar{\beta}^1 > f_1(x^1)$, $\bar{x}^2 \in X_2$, $\bar{\beta}^2 < f_2(x^2)$. For $\lambda = 0$, these relations mean that the sets X_1 and X_2 are properly separated and $\operatorname{ri} X_1 \cap \operatorname{ri} X_2 = \varnothing$. But this contradicts the condition of this theorem. That is why $\lambda \neq 0$. At the same time, from [3.23], as $\beta^1 \to +\infty$ (or $\beta^2 \to -\infty$) it follows that $\lambda \geqslant 0$. Therefore $\lambda > 0$. Take $a = -p/\lambda$, $b = \alpha/\lambda$. Note that the strict inequalities in [3.23] hold true for $\beta^1 = f_1(x^1)$, $\beta^2 = f_2(x^2)$. Considering such β^1 i β^2, we get

$$f_1(x^1) - \langle a, x^1 \rangle \geqslant b \geqslant f_2(x^2) - \langle a, x^2 \rangle$$

for all $x^1 \in X_1$, $x^2 \in X_2$. In other words, the linear function $l(x) = \langle a, x \rangle + b$ satisfies conditions [3.21] and [3.22]. □

REMARK 3.5.– Condition $\operatorname{ri} X_1 \cap \operatorname{ri} X_2 \neq \varnothing$ in the theorem is essential. This can be concluded from the example: $X_1 = [-1, 1]$, $X_2 = [1, \infty]$, $f_1(x) = -\sqrt{1 - x^2}$, $f_2(x) = 0$.

3.5. Convex minimization problem

THEOREM 3.23.– Let X be a convex set and let $f(x)$ be a convex function on X. Then the local solution to the minimization problem

$$f(x) \to \min, \quad x \in X,$$

is the global solution to this minimization problem.

PROOF.– Let $\hat{x}$ be a local solution to the minimization problem. Then for some $\varepsilon > 0$, the following inequality holds true:

$$f(\hat{x}) \leq f(x) \quad \text{for all} \quad x \in X \cap B_\varepsilon(\hat{x}).$$

For any point $x \in X, x \neq \hat{x}$, take $\lambda = \min\{\varepsilon/\|x - \hat{x}\|, 1\}$. Then $\lambda x + (1-\lambda)\hat{x} \in X \cap B_\varepsilon(\hat{x})$ and

$$f(\hat{x}) \leq f(\lambda x + (1 - \lambda)\hat{x}) \leq \lambda f(x) + (1 - \lambda) f(\hat{x}).$$

Therefore

$$f(\hat{x}) \leq f(x) \quad \text{for all} \quad x \in X.$$

This means that $\hat{x}$ is the global solution to the minimization problem. □

Consequently, for convex minimization problems, the concepts of local and global solutions do not differ and we can talk simply about the solution of the minimization problem.

Another important property of convex minimization problems can be formulated in the form of the following general principle: the necessary conditions for optimality in one or another class of optimization problems with the corresponding assumptions of convexity are sufficient conditions for optimality. As an example, let us formulate the following theorem.

THEOREM 3.24.– Let a function f be convex on $\mathbb{R}^n$ and differentiable at a point $\hat{x} \in \mathbb{R}^n$. If $f'(\hat{x}) = 0$, then $\hat{x}$ is a point of minimum of the function $f(x)$.

PROOF.– For all $x \in \mathbb{R}^n$ and $\lambda \in (0, 1]$, we have

$$f(\lambda x + (1 - \lambda)\hat{x}) \leq \lambda f(x) + (1 - \lambda) f(\hat{x}).$$

Using the fact that the function f is differentiable at the point $\hat{x}$, we have

$$f(x) - f(\hat{x}) \geq \frac{f(\hat{x} + \lambda(x - \hat{x})) - f(\hat{x})}{\lambda}$$

$$= \frac{\langle f'(\hat{x}), \lambda(x - \hat{x})\rangle + o(\lambda)}{\lambda} = \frac{o(\lambda)}{\lambda}.$$

Passing on to the limit as $\lambda \to 0$, we will have $f(x) \geq f(\hat{x})$. This means that $\hat{x}$ is the global solution to this minimization problem. □

We give here another property of convex minimization problems.

THEOREM 3.25.– Let X be a convex set and let $f(x)$ be a convex function on X. Then the set of (global) solutions $\hat{X} = \operatorname{Arg\,min}_{x \in X} f(x)$ of the minimization problem is convex. If a function $f(x)$ is strictly convex on X, then solution to the minimization problem is unique (the set $\hat{X} = \operatorname{Arg\,min}_{x \in X} f(x)$ contains only one point).

PROOF.– Let $x^1, x^2 \in \hat{X}, \lambda \in [0, 1]$. Then $f(x^1) = f(x^2) = \hat{f}$ and the following inequality holds true:

$$f(\lambda x^1 + (1 - \lambda)x^2) \leq \lambda f(x^1) + (1 - \lambda) f(x^2) = \hat{f}.$$

By definition of $\hat{X}$, this inequality holds true as an equality only. That is why $\lambda x^1 + (1 - \lambda)x^2 \in \hat{X}$, i.e. $\hat{X}$ is a convex set.

Let the function f be strictly convex on X. Assuming that there are two distinct points $x^1, x^2, x^1 \neq x^2$ in $\hat{X}$, then for $\lambda \in (0, 1)$ in the last relation the inequality must be strict, that is impossible. □

3.6. Theorem on boundedness of Lebesgue set of a strongly convex function

Let X be a convex set and let $f(x)$ be a convex function on X. The set

$$X_\beta = \{x \in X \mid f(x) \leqslant \beta\}, \quad \beta \in \mathbb{R} \qquad [3.24]$$

in this case is convex. This set is called the *Lebesgue set* (or *sublevel set*) of the function f. In the following theorem, one important property of the Lebesgue set of a strongly convex function is described.

THEOREM 3.26.– Let $f(x)$ be a continuous strongly convex function with the module of convexity $\theta > 0$ on a closed convex set X. Then for arbitrary β the set X_β is bounded.

PROOF.– Let $x^0 \in X_\beta$ be a fixed point (if $X_\beta = \varnothing$, then the statement is trivial). Let $U = U_1(x^0)$ be a sphere of the unit radius with the center at point x^0. It follows from the continuity of the function f and closedness of the set X that there exists a constant α such that

$$f(x) \geqslant \alpha \quad \text{for all} \quad x \in X \cap U. \qquad [3.25]$$

We will show that

$$\|x - x^0\| \leqslant 1 + \frac{\beta - \alpha}{\theta} \quad \text{for all} \quad x \in X_\beta. \qquad [3.26]$$

This means that the set X_β is bounded. If $x \in X_\beta \cap U$, then $\|x - x^0\| \leqslant 1$. Inequality [3.26] holds true. Let $x \in X_\beta \backslash U$. Take $\lambda = 1/\|x - x^0\|$, $\bar{x} = \lambda x + (1 - \lambda)x^0$. Then $0 < \lambda < 1$ and $\bar{x} \in X \cap U$. Making use of the inequality [3.25], the strong convexity of the function f, conditions $f(x^0) \leqslant \beta$, $f(x) \leqslant \beta$ and definition of λ, we get

$$\alpha \leqslant f(\bar{x}) \leqslant \lambda f(x) + (1 - \lambda) f(x^0) - \theta\lambda(1 - \lambda)\|x - x^0\|^2$$
$$\leqslant \beta - \theta\lambda(1 - \lambda)\|x - x^0\|^2 = \beta - \theta(\|x - x^0\| - 1)$$

From here, we come to [3.26]. □

COROLLARY 3.3.– Under conditions of theorem 3.26, a point of minimum of the function f on the set X exists and it is unique.

Note that the set X in this theorem is not necessary bounded. We can take, for example, $X = \mathbb{R}^n$. Statements analogous to assertions of theorem 3.26 and corollary 3.3 for convex and strictly convex functions are not true. As an example, we can consider the function $f(x) = e^x$ on $\mathbb{R}$. However, for convex functions the following analogue of theorem 3.26 holds true.

THEOREM 3.27.– Let $f(x)$ be a continuous convex function on a closed convex set X. Suppose that for some $\bar{\beta}$, the set $X_{\bar{\beta}}$ of the form [3.24] is non-empty and bounded. Then for all β, the set X_β is bounded.

PROOF.– If $\beta < \bar{\beta}$, then $X_\beta \subset X_{\bar{\beta}}$ and the boundedness of X_β is obvious. Let $\beta > \bar{\beta}$. Suppose that X_β is unbounded. Consider an arbitrary point $\hat{x} \in X_{\bar{\beta}} \subset X_\beta$. Since X_β is convex and closed, then there exists a ray which starts from the point $\hat{x}$ that entirely lies in X_β. In other words, there exists a vector $h \neq 0$ such that

$$\hat{x} + \alpha h \in X_\beta, \quad \text{that is} \quad \hat{x} + \alpha h \in X \quad \text{and} \quad f(\hat{x} + \alpha h) \leqslant \beta \qquad [3.27]$$

for all $\alpha \geqslant 0$. Define the function $\varphi(\alpha) = f(\hat{x} + \alpha h)$ on $\mathbb{R}_+$. Suppose that $\varphi(\bar{\alpha}) > \varphi(0)$ for some $\bar{\alpha} > 0$. By theorem 3.12, the function φ is convex on $\mathbb{R}_+$. Then for all $\alpha > \bar{\alpha}$, we have

$$\varphi(\bar{\alpha}) = \varphi\left(\left(1 - \frac{\bar{\alpha}}{\alpha}\right) \cdot 0 + \frac{\bar{\alpha}}{\alpha}\alpha\right) \leqslant \left(1 - \frac{\bar{\alpha}}{\alpha}\right)\varphi(0) + \frac{\bar{\alpha}}{\alpha}\varphi(\alpha).$$

Therefore

$$\varphi(\alpha) \geqslant \frac{\alpha}{\bar{\alpha}}\left(\varphi(\bar{\alpha}) - \varphi(0)\right) + \varphi(0).$$

Since $\varphi(\bar{\alpha}) > \varphi(0)$, then $\varphi(\alpha) \to +\infty$ as $\alpha \to +\infty$, which contradicts [3.27]. So, for each $\alpha > 0$

$$f(\hat{x} + \alpha h) = \varphi(\alpha) \leqslant \varphi(0) = f(\hat{x}) \leqslant \bar{\beta},$$

that is $\hat{x} + \alpha h \in X_{\bar{\beta}}$. But this contradicts the boundedness of $X_{\bar{\beta}}$. □

COROLLARY 3.4.– Let $f_1, \ldots, f_m$ be continuous convex functions on a closed convex set X. Consider sets of the form

$$X(b) = \{x \in X \mid f_i(x) \leqslant b_i, \, i = 1, \ldots, m\},$$

where $b = (b_1, \ldots, b_m) \in \mathbb{R}^m$. Suppose that for some $\bar{b}$, the set $X(\bar{b})$ is non-empty and bounded. Then for all b the set $X(b)$ is bounded.

PROOF.– For the function $f(x) = \max\limits_{1 \leq i \leq m} \left(f_i(x) - \bar{b}_i\right)$, consider sets X_β of the form [3.24]. Under the conditions of the theorem, the set $X_0 = X\left(\bar{b}\right)$ is non-empty and bounded. The function f is convex on X (theorem 3.9) and continuous. For every b there is β, such that $X(b) \subset X_\beta$. By theorem 3.27, the set X_β is bounded. Therefore, the set $X(b)$ is bounded. □

One important property of the convex optimization problem follows from these statements.

THEOREM 3.28.– Consider the convex optimization problem

$$\begin{aligned} &f(x) \to \min, \\ &g_i(x) \leqslant b_i, \quad i = 1, \dots, m, \\ &x \in P, \end{aligned} \qquad [3.28]$$

where P is a closed convex set in $\mathbb{R}^n$, and $f(x)$, $g_1(x), \dots, g_m(x)$ are continuous convex functions on P. For every $b = (b_1, \dots, b_m) \in \mathbb{R}^m$, we denote by $X(b)$ and $X^*(b)$, respectively, the admissible set and the set of solutions to the problem [3.28]. Suppose that for some $\bar{b}$, the set $X^*(\bar{b})$ is non-empty and bounded. Then for all b such that $X(b) \neq \varnothing$, the set $X^*(b)$ is non-empty and bounded.

PROOF.– Consider a set of the form

$$Y(b, \beta) = \{x \in X(b) \mid f(x) \leqslant \beta\},$$

where $b \in \mathbb{R}^m$, $\beta \in \mathbb{R}$. Take $\bar{\beta} = f(x)$ for $x \in X^*(\bar{b})$. Under conditions of the theorem, the set $Y(b, \bar{\beta}) = X^*(b)$ is non-empty and bounded. Consider an arbitrary vector $b \in \mathbb{R}^m$, such that there exists $x^0 \in X(b)$. By corollary 3.4, the non-empty set $Y(b, f(x^0))$ is bounded. Then the set $X^*(b)$ is non-empty, by the Weierstrass theorem, and bounded, since $X^*(b) \subset Y(b, f(x^0))$. □

3.7. Conjugate function

DEFINITION 3.12.– *Let a function* f: $X \to \mathbb{R}$, $X \subset \mathbb{R}^n$. *The function* f^*: $Y \to \mathbb{R}$, $Y \subset \mathbb{R}^n$, *defined as*

$$f^*(y) = \sup_{x \in X} \left(\langle y, x \rangle - f(x) \right),$$

is called the conjugate function *to the function* f. *The domain of definition of the function* f^* *is the set*

$$Y = \left\{ y \in \mathbb{R}^n \;\middle|\; \sup_{x \in X} \left(\langle y, x \rangle - f(x) \right) < \infty \right\}.$$

REMARK 3.6.– Note that the function f^* is convex, since it is the supremum of a family on the convex function with respect to y (irrespective of whether the function f itself is convex).

Let us find conjugate functions to some convex functions:

– *Affine function* $f(x) = ax + b$: the function $yx - ax - b$ (with respect to x) is bounded if and only if $y = a$. In this case, this function is constant. The domain of definition of the conjugate function f^* is the set $Y = \{a\}$ and $f^*(a) = -b$.

– *Negative logarithmic function* $f(x) = -\ln x$ with the domain of definition $\operatorname{int}\mathbb{R}_+$: the function $xy + \ln x$ is unbounded from above when $y \geqslant 0$ and attains its maximum at point $x = -1/y$ when $y < 0$. Therefore, the conjugate function f^* is defined on the set $Y = \{y \mid y < 0\}$ and $f^*(y) = -\ln(-y) - 1$ for $y \in Y$.

– *Exponential function* $f(x) = e^x$: the function $xy - e^x$ is unbounded when $y \leqslant 0$. For $y > 0$, the function $xy - e^x$ attains its maximum at point $x = \ln y$. Therefore, $f^*(y) = y\ln y - y$ for $y > 0$, and $f^*(y) = \sup_x(-e^x) = 0$ for $y = 0$. Therefore, the conjugate function f^* is defined on $\mathbb{R}_+$ and $f^*(y) = y\ln y - y$ (here $0\ln 0 = 0$).

– *Negative entropy function* $f(x) = x\ln x$ defined on $\mathbb{R}_+$ ($f(0) = 0$): the function $xy - x\ln x$ is bounded from above on $\mathbb{R}_+$ for all y. Therefore, the conjugate function f^* is defined on $\mathbb{R}$ and attains its maximum at point $x = e^{y-1}$. Therefore, the conjugate function is $f^*(y) = e^{y-1}$.

– *Inverse function* $f(x) = 1/x$ on $\mathbb{R}_+$. For $y > 0$, the function $yx - 1/x$ is unbounded from above. For $y = 0$, the supremum of the function is equal to 0. For $y < 0$, the supremum is attained at point $x = (-y)^{-1/2}$: Therefore, the conjugate function $f^*(y) = -2(-y)^{1/2}$ is defined on $\mathbb{R}_-$.

– *Strictly convex quadratic form*: consider the function $f(x) = \frac{1}{2}\langle Qx, x\rangle$, where $Q \in \mathbb{S}^n_+$. The function $\langle x, y\rangle - \frac{1}{2}\langle Qx, x\rangle$ is bounded from above as a function of x for all y. It attains its maximum at point $x = Q^{-1}y$. Therefore, the conjugate function

$$f^*(y) = \frac{1}{2}\langle Q^{-1}y, y\rangle.$$

– *Logarithm of determinant of a matrix*: the function $f(X) = \ln\det X^{-1}$ is determined on the set of positive definite matrices. The conjugate function is defined as

$$f^*(Y) = \sup_{X>0}\left(\operatorname{Tr} XY + \ln\det X\right).$$

The expression under the sign of the supremum is unbounded from above if the matrix Y is not negative definite. In the case $Y < 0$, we can find the maximum by equating to zero the gradient of the expression with respect to X:

$$(\operatorname{Tr} XY + \ln\det X)'_X = Y + X^{-1} = 0,$$

We get $X = -Y^{-1}$. The conjugate function is defined as

$$f^*(y) = \ln\det(-Y)^{-1} - n$$

on the set of negative definite matrices.

– *Indicator function of the set*: let $\delta(x|C)$ be the indicator function of a (not necessarily convex) set $C \subset \mathbb{R}^n$, that is $\delta(x|C) = 0$ on C. The conjugate function to the indicator function is the support function to the set C:

$$\delta^*(x|C) = \sup_{x \in C}\langle y, x\rangle = \sigma(x|C).$$

– *Logarithm of the sum of exponents*: to find the conjugate function to the function $f(x) = \ln \sum_{i=1}^{n} e^{x_i}$, we have to find y for which the function $\langle x, y\rangle - f(x)$ attains its maximum with respect to x. Taking the derivative with respect to x of the function equal to zero, we get the following equations:

$$y_i = \frac{e^{x_i}}{\sum_{j=1}^{n} e^{x_j}}, \qquad i = 1, \ldots, n.$$

These equations have a solution with respect to x if and only if $y_i > 0$ and $\sum_{i=1}^{n} y_i = 1$. Substituting expressions for y_i in $\langle x, y\rangle - f(x)$, we obtain $f^*(y) = \sum_i y_i \ln y_i$. This function will remain to be defined in the case where we set some components y as equal to 0 ($0 \ln 0$ we assume equal to 0). So the domain of definition of the function f^* is determined by the relations $y_i \geqslant 0$, $i = 1, \ldots, n$, $\sum_{i=1}^{n} y_i = 1$. Suppose that there exists k, such that $y_k < 0$. Show that in this case the function $\langle x, y\rangle - f(x)$ is unbounded from above. Take $x_k = -t$ and $x_i = 0$ for $i \neq k$ and direct t to infinity. If $y_i \geqslant 0$, but $\sum_{i=1}^{n} y_i = 1$, take $x = t\,\mathbb{I}$. Hence

$$\langle x, y\rangle - f(x) = t\langle \mathbb{I}, y\rangle - t - \ln n.$$

This expression goes to infinity as $t \to \infty$. This is why

$$f^*(y) = \begin{cases} \sum_{i=1}^{n} y_i \ln y_i, & \text{if } y_i \geqslant 0,\ \sum_{i=1}^{n} y_i = 1, \\ +\infty, & \text{in other cases.} \end{cases}$$

– *Norm*: let $\|\cdot\|$ be a norm in $\mathbb{R}^n$ with the conjugate norm $\|\cdot\|_*$. We will show that the conjugate function to the function $f(x) = \|x\|$ is the function

$$f^*(y) = \begin{cases} 0, & \|y\|_* \leqslant 1; \\ \infty, & \text{in other cases.} \end{cases}$$

In other words, the conjugate function is the indicator function of a unit sphere in the conjugate space. If $\|y\|_* > 1$, then by definition of the conjugate function there exists $z \in \mathbb{R}^n$ such that $\|z\| \leqslant 1$ and $\langle y, z\rangle > 1$. When we take $x = tz$ and direct $t \to \infty$, we have

$$\langle y, x\rangle - \|x\| = t(\langle y, z\rangle - \|z\|) \to \infty.$$

This shows that $f^*(y) = \infty$. On the contrary, if $\|y\|_* \leqslant 1$, then, since $\langle y, x\rangle \leqslant \|x\|\|y\|_*$ for all x, we have $\langle y, x\rangle - \|x\| \leqslant 0$. Thus $x = 0$ is the point at which the maximum of the expression $\langle y, x\rangle - \|x\|$ is attained, and this maximum is equal to zero.

– *Square of norm*: consider the function $f(x) = \frac{1}{2}\|x\|^2$. We show that the conjugate to this function is the function $f^*(y) = \frac{1}{2}\|y\|_*^2$. Since $\langle y, x\rangle \leqslant \|y\|_*\|x\|$, we can conclude that

$$\langle y, x\rangle - \frac{1}{2}\|x\|^2 \leqslant \|y\|_*\|x\| - \frac{1}{2}\|x\|^2$$

for all x. The expression in the right side is a quadratic function of $\|x\|$. The maximum value of this function is $(1/2)\|y\|_*^2$. That is why for all x we have

$$\langle y, x\rangle - (1/2)\|x\|^2 \leqslant (1/2)\|y\|_*^2.$$

The last inequality shows that $f^*(y) \leqslant (1/2)\|y\|_*^2$.

To prove the inequality for the other side, we choose x to be such a vector that $\langle y, x\rangle = \|y\|_*\|x\|$, and $\|x\| = \|y\|_*$. For this x, we have

$$\langle y, x\rangle - (1/2)\|x\|^2 = \|y\|_*\|x\| - (1/2)\|y\|_* = (1/2)\|y\|_*^2.$$

Hence $f^*(y) = (1/2)\|y\|_*^2$.

– *Profit function*: consider a firm, which uses n resources to produce a product that can sell. Denote by $r = (r_1, \ldots, r_n)$ the vector that defines the quantity of resources used. Denote by $S(r)$ the profit from the sale as a function of the resources used. By p_i, we denote the price per unit of the ith resource. The total amount the firm spent on resources is $\langle p, r\rangle$. The profit earned by the firm is $S(r) - \langle p, r\rangle$. We fix the prices of resources and determine the maximum profit the firm can get, correctly choosing the number of used resources. The largest profit equals

$$M(p) = \sup_r \left(S(r) - \langle p, r\rangle\right).$$

$M(p)$ is the maximum profit that a firm can get depending on resource prices. Using conjugation, we can represent M in the form

$$M(p) = (-S)^*(-p).$$

Consequently, the maximum profit (as a function of the prices of resources) is tightly related to gross sales (as a function of used resources).

3.8. Basic properties of conjugate functions

– *Fenchel's inequality*: from the definition of a conjugate function, it follows that

$$f(x) + f^*(y) \geqslant \langle x, y\rangle \quad \text{for all} \quad x,\ y. \qquad [3.29]$$

This inequality is called *Fenchel's inequality* (Young's inequality for differentiable functions).

For the function $f(x) = (1/2)\langle Qx, x\rangle$, where Q is a positive definite matrix, we get the inequality

$$\langle x, y\rangle \leqslant \frac{1}{2}\langle Qx, x\rangle + \frac{1}{2}\langle Q^{-1}y, y\rangle.$$

– *Conjugate to conjugate function*: the above examples show that the function conjugate to the conjugate function is the function f itself.

This assertion holds in the case where the function f is convex and its epigraph epi $f = \{(x, y) \mid f(x) \leqslant y\}$ is a closed set. For example, if the function f is convex on $\mathbb{R}^n$, then $f^{**} = f$.

This means that if the function f is convex and its epigraph is a closed set, then for every x there exists a y such that inequality [3.29] is transformed into equality.

– *Conjugate to differentiable function*: the conjugate to differentiable function f is also called the *Legendre transformation* of the function f.

Let f be a convex and differentiable function on $\mathbb{R}^n$. Every x^*, at which the maximum of the function $\langle x, y\rangle - f(x)$ is achieved, satisfies the condition $y = f'(x^*)$. Conversely, if $y = f'(x^*)$, then the maximum of $\langle x, y\rangle - f(x)$ is achieved at x^*. So, if $y = f'(x^*)$, then

$$f^*(y) = \langle x^*, f'(x^*)\rangle - f(x^*).$$

This allows us to find $f^*(y)$ at those points for which we can solve the equation $y = f'(z)$.

– For $a > 0$ and $b \in \mathbb{R}$, the conjugate to the function $g(x) = af(x) + b$ is the function $g^*(y) = af^*(y/a) - b$.

– Let A be a non-degenerate matrix of dimension $n \times n$, and let $b \in \mathbb{R}^n$. The conjugate to function $g(x) = f(Ax + b)$ is the function

$$g^*(y) - f^*\left(\left(A^{-1}\right)^{\top} y\right) - b^{\top}\left(A^{-1}\right)^{\top} y.$$

– *Infimal convolution of functions*: let the functions $f_1 \colon \mathbb{R}^n \to \mathbb{R}$ and $f_2 \colon \mathbb{R}^n \to \mathbb{R}$ have the conjugate functions f_1^* and f_2^* correspondingly. Then the conjugate function to the infimal convolution of the functions f_1 and f_2 is equal to the sum of the conjugate functions f_1^* and f_2^*:

$$(f_1 \oplus f_2)^*(y) = f_1^*(y) + f_2^*(y).$$

Actually

$$(f_1 \oplus f_2)^*(y) = \sup_x \left\{\langle x, y\rangle - \inf_{x_1 + x_2 = x} \{f_1(x_1) + f_2(x_2)\}\right\}$$

$$= \sup_x \sup_{x_1+x_2=x} \{\langle x, y\rangle - f_1(x_1) - f_2(x_2)\}$$

$$= \sup_{x_1,x_2} \{[\langle x_1, y\rangle - f_1(x_1)] + [\langle x_2, y\rangle - f_2(x_2)]\}$$

$$= f_1^*(y) + f_2^*(y).$$

– *The sum of independent functions*: if $f(u,v) = f_1(u) + f_2(v)$, where f_1 and f_2 are convex functions with the conjugate functions f_1^* and f_2^* correspondingly, then

$$f^*(w,z) = f_1^*(w) + f_2^*(z).$$

In other words, the conjugate of a sum of independent functions is a sum of conjugated functions. Here, under independent functions, we understand the functions of different variables.

3.9. Exercises

1) Let f be a convex function on a convex set X. Prove that

$$f(\lambda x^1 + (1-\lambda)x^2) \geqslant \lambda f(x^1) + (1-\lambda)f(x^2)$$

for all x^1, $x^2 \in X$, $\lambda \notin [0,1]$, such that $\lambda x^1 + (1-\lambda)x^2 \in X$.

2) Let f be a strictly convex function on a convex set X. Prove that

$$f(\lambda x^1 + (1-\lambda)x^2) > \lambda f(x^1) + (1-\lambda)f(x^2)$$

for all x^1, $x^2 \in X$, $x^1 \neq x^2$, $\lambda \notin [0,1]$, such that $\lambda x^1 + (1-\lambda)x^2 \in X$.

3) Let f be a strongly convex function on a convex set X. Prove that

$$f(\lambda x^1 + (1-\lambda)x^2) \geqslant \lambda f(x^1) + (1-\lambda)f(x^2) - \theta\lambda(1-\lambda)\left\|x^1 - x^2\right\|^2$$

for all x^1, $x^2 \in X$, $\lambda \notin [0,1]$, such that $\lambda x^1 + (1-\lambda)x^2 \in X$.

4) Let f be a continuous midpoint convex function on a convex set X, that is, for all $x_1 \in X$, $x_2 \in X$

$$f\left(\frac{x_1+x_2}{2}\right) \leq \frac{f(x_1)+f(x_2)}{2}.$$

Prove that the function f is convex on X.

5) Let f be a continuous function on a convex set X, and let, for all points x^1, $x^2 \in X$, the following inequality holds:

$$f\left(\frac{x_1+x_2}{2}\right) \leq \frac{f(x_1)+f(x_2)}{2} - \theta\|x_1 - x_2\|^2,$$

where $\theta > 0$ is the same for all x^1, $x^2 \in X$. Prove that the function f is strongly convex on X.

6) Let f be a continuous function on a convex set X, and let, for all points $x^1, x^2 \in X$, there exists a number $\lambda \in (0, 1)$, such that

$$f(\lambda x^1 + (1 - \lambda)x^2) \leqslant \lambda f(x^1) + (1 - \lambda)f(x^2).$$

Prove that the function f is convex on X.

7) Let $\lambda_1 \geqslant 0, \dots, \lambda_m \geqslant 0$, $\sum_{i=1}^{m} \lambda_i = 1$. Applying the Jensen inequality to the corresponding functions, prove the following inequalities:

a) $\left(\sum\limits_{i=1}^{m} \lambda_i x_i\right)\left(\sum\limits_{i=1}^{m} \frac{\lambda_i}{x_i}\right) \geqslant 1$, where $x_1 > 0, \dots, x_m > 0$;

b) $\sum\limits_{i=1}^{m} \lambda_i x^i \leqslant \ln\left(\sum\limits_{i=1}^{m} \lambda_i e^{x_i}\right)$, where $x_1, \dots, x_m$ are arbitrary numbers.

8) Let φ be a convex function on a convex set X. Check the following statements:

a) the function $f(x) = e^{\varphi(x)}$ is convex on X;

b) the function $f(x) = 1/\varphi(x)$ is concave on $X_0 = \{x \in X \mid \varphi(x) < 0\}$.

9) Let $a_i \in \operatorname{int} \mathbb{R}^n_+$, $i = 1, \dots, m$. Prove that the function

$$f(x) = \sum_{i=1}^{m} \frac{1}{\langle a_i, x\rangle}$$

is convex on $\mathbb{R}^n_+ \setminus \{0\}$.

10) Let X be a convex set. Prove the following statements:

a) If $f_i(x), i = 1, \dots$ are convex uniformly bounded functions on X, and $\alpha_i, i = 1, \dots$ are non-negative numbers such that $\sum_{i=1}^{\infty} \alpha_i < \infty$, then the function

$$f(x) = \sum_{i=1}^{\infty} \alpha_i f_i(x)$$

is convex on X.

b) Let $\varphi(x, t)$ be a function on $X \times [0, 1]$, such that it is convex with respect to x on X for all $t \in [0, 1]$ and integrable with respect to t on $[0, 1]$ for all $x \in X$, and let $\alpha(t)$ be a non-negative integrable function on $[0, 1]$, then the function

$$f(x) = \int_0^1 \varphi(x, t)\alpha(t)dt$$

is convex on X.

11) a) Prove that in theorem 3.18 for $\theta = 0$ (case of convex function), it is sufficient to assume that f is differentiable on X only. Hint: use the Lagrange formula instead of the Newton–Leibniz formula.

b) Ensure that the continuity of the matrix $f''(x)$ in theorem 3.19 is necessary only for the derivation of the formula [3.15] for points $\hat{x} \in X \setminus \operatorname{int} X$.

c) Make sure that the inverse statement of theorem 3.19 admits the case $\operatorname{int} X = \varnothing$; give an example that shows that in the direct statement of theorem 3.19 the condition $\operatorname{int} X \neq \varnothing$ is essential.

12) Let f be a two times differentiable function on a convex compactum X, and $\langle f''(\hat{x})h, h\rangle > 0$ for all $\hat{x} \in X$, $h \in \mathbb{R}^n$, $h \neq 0$. Prove that the function f is strongly convex on X.

13) Which of the following functions is convex, concave or neither and why?

a) $f(x_1, x_2) = 2x_1 - 4x_1x_2 - 8x_1 + 3x_2$,

b) $f(x_1, x_2) = x_1 \exp(-x_1 - 3x_2)$,

c) $f(x_1, x_2) = 2x_1 + 6x_2 - 2x_1^2 - 3x_2^2 + 4x_1x_2$,

d) $f(x_1, x_2, x_3) = 2x_1x_2 + 2x_1^2 + x_2^2 + 2x_3^2 - 5x_1x_3$,

14) Over what domain is the function $f(x) = x^2(x^2 - 1)$ convex? Is it strictly convex over the regions specified?

15) Check that the function $f(x) = \sqrt{1 + x_1^2 + x_2^2}$ is convex on $\mathbb{R}^2$.

16) Let φ be a continuous monotonically non-decreasing (increasing) function on the interval $[a, b]$. Prove that the function $f(x) = \int_a^x \varphi(t)dt$ is convex (strictly convex) on $[a, b]$.

17) Let the function f be convex on $\mathbb{R}$. Prove that the function

$$F(x) = (1/x) \int_0^x f(t)dt, \quad F(0) = 0, \quad x \geq 0,$$

is convex.

18) a) Let f be a differentiable strictly convex function on a convex set $X \subset \mathbb{R}^n$. Show that for arbitrary $a \in \mathbb{R}^n$, the equation $f'(x) = a$ has no more than one solution on X.

b) Let f be a differentiable strongly convex function on $\mathbb{R}^n$. Show that for arbitrary $a \in \mathbb{R}^n$ a solution of the equation $f'(x) = a$ on $\mathbb{R}^n$ exists and it is unique.

19) a) Let f be a differentiable convex function on $\mathbb{R}^n$. Show that for arbitrary $\lambda > 0$ a solution of the equation $f'(x) = -\lambda x$ on $\mathbb{R}^n$ exists and it is unique.

b) Show that the system of equations

$$\lambda x_1 + 2e^{2x_1+3x_2} + 2(x_1 - x_2)e^{(x_1-x_2)^2} = 0$$

$$\lambda x_2 + 3e^{2x_1+3x_2} + 2(x_1 - x_2)e^{(x_1-x_2)^2} = 0$$

has a unique solution for arbitrary $\lambda > 0$.

20) Let f be a convex function on a convex set $X \subset \mathbb{R}^n$. Show that the function $f'(x; h)$ on $\operatorname{ri} X \times \operatorname{Lin} X$ has the following properties:

a) $f'(x; h)$ is convex with respect to h on $\operatorname{Lin} X \quad \forall x \in \operatorname{ri} X$;

b) $f'(x; h) \geqslant -f'(x; -h) \quad \forall h \in \operatorname{Lin} X, \forall x \in \operatorname{ri} X$;

c) $f'(x; h)$ is upper semicontinuous on $\operatorname{ri} X \times \operatorname{Lin} X$.

21) Let f be a strongly convex function with the constant $\theta > 0$ on a convex set X, and let $\hat{x}$ be a point of minimum of the function f on X. Prove the following estimates:

a) $\theta\|x - \hat{x}\|^2 \leqslant f(x) - f(\hat{x}) \ \forall x \in X$;

b) if f is differentiable at point $x \in X$, then

$$2\theta\|x - \hat{x}\| \leqslant \|f'(x)\|$$

and

$$4\theta(f(x) - f(\hat{x})) \leqslant \|f'(x)\|^2.$$

22) Let f be a bounded from above convex function on $\mathbb{R}^n$ (on $\mathbb{R}^n_+$). Show that f is a constant function on $\mathbb{R}^n$ (f is monotonically nonincreasing on $\mathbb{R}^n_+$).

23) Find conjugate functions to the following functions:

a) *Maximum*: $f(x) = \max\limits_{1 \leqslant i \leqslant n} \{x_i\}$ on $\mathbb{R}^n$.

b) *Sum of the largest components*: $f(x) = \sum_{i=1}^{r} x_{[i]}$ on $\mathbb{R}^n$.

c) *Piecewise linear function* $f(x) = \max\limits_{i=1,\dots,m} (\langle a_i, x\rangle + b_i)$.

d) *Power function* $f(x) = x^p$ on $\operatorname{int} \mathbb{R}^n_+$ for $p > 1$ on $p < 0$.

e) *Negative geometric mean* $f(x) = -\left(\prod x_i\right)^{1/n}$ on $\mathbb{R}^n_+$.

24) Show that the conjugate function to the function $f(X) = \operatorname{Tr}(X^{-1})$ on the set of positive definite matrices is the function

$$f^*(Y) = -2\operatorname{Tr}(-Y)^{1/2},$$

which is defined on the set of negative definite symmetric matrices.

25) Let $g(x) = f(x) + \langle c, x\rangle + d$, where f is a convex function. Express g^* in terms of f^*.

26) *Derivatives from conjugate function.* Let $f\colon \mathbb{R}^n \to \mathbb{R}$ be a convex and two times differentiable function on $\mathbb{R}$. Let $\tilde{y}$ and $\tilde{x}$ be such that $\tilde{y} = f'(\tilde{x})$, and let the derivative $f'(x)$ have an inverse in the neighborhood of $\tilde{x}$. In other words for each y in the neighbourhood of $\tilde{y}$, there exists unique x in the neighborhood of $\tilde{x}$ such that $y = f'(x)$. Show that the following relations hold true:

a) $(f^*)'(\tilde{y}) = \tilde{x}$,

b) $(f^*)''(\tilde{y})f''(\tilde{x}) = I$.

27) *Young's inequality.* Let f be an increasing function on $\mathbb{R}$, $f(0) = 0$, and let g be the function inverse to f. Define F and G in the following way:

$$F(x) = \int_0^x f(a)da, \quad G(y) = \int_0^y g(a)da.$$

Prove that the functions F and G are conjugate. Give a geometric interpretation of Young's inequality

$$xy \leqslant F(x) + G(y).$$

a) Show that the necessary condition of convexity of two times differentiable functions:

$$\langle y, f'(x)\rangle = 0 \Rightarrow \langle f''(x)y, y\rangle \geqslant 0 \quad \text{for all} \quad x \in X.$$

can be expressed in the following two equivalent ways:

– for all $x \in X$, there exists $\lambda(x) \geqslant 0$ such that

$$f''(x) + \lambda(x)f'(x)\left(f'(x)\right)^{\top} \geqslant 0;$$

– for all $x \in X$, the matrix

$$\begin{pmatrix} f''(x) & f'(x) \\ (f'(x))^{\top} & 0 \end{pmatrix}$$

has no more than one negative eigenvalue.

b) Show that the sufficient condition of convexity of two times differentiable functions

$$\langle y, f'(x)\rangle = 0 \Rightarrow \langle f''(x)y, y\rangle > 0 \quad \text{for all} \quad x \in X, y \in \mathbb{R}^n.$$

can be expressed in the following two equivalent ways:

– for all $x \in X$, there exists $\lambda(x) \geqslant 0$ such that

$$f''(x) + \lambda(x)f'(x)\left(f'(x)\right)^\top > 0;$$

– for all $x \in X$, the matrix

$$\begin{pmatrix} f''(x) & f'(x) \\ (f'(x))^\top & 0 \end{pmatrix}$$

has one non-negative and n positive eigenvalues.

4

Generalizations of Convex Functions

4.1. Quasi-convex functions

DEFINITION 4.1.– *Let X be a convex subset of $\mathbb{R}^n$. A function $f\colon X \to \mathbb{R}$ is called* quasi-convex *(or* unimodal*) if all its Lebesgue sets*

$$X_\beta = \{x \in X \mid f(x) \leqslant \beta\}$$

are convex. A function f is called quasi-concave *if the function $g = -f$ is quasi-convex. A function which is simultaneously quasi-convex and quasi-concave is called* quasi-linear.

Convex functions have convex Lebesgue sets. So convex functions are quasi-convex one. The inverse statement is not true.

EXAMPLE 4.1.– Examples of quasi-convex functions on $\mathbb{R}$ are as follows:

– the function $f(x) = \ln x$ on $\operatorname{int}\mathbb{R}_+$ is quasi-convex (and quasi-concave, so it is quasi-linear);

– the function $f(x) = \operatorname{ceil}(x) = \min\{z \in \mathbb{Z} \mid z \geqslant x\}$ is quasi-convex (and quasi-concave).

These examples show that a quasi-convex function can be concave, even discontinuous.

Examples of quasi-convex functions on $\mathbb{R}^n$ are as follows.

EXAMPLE 4.2.– *(Length of the vector)* Define the length of a vector $x \in \mathbb{R}^n$ as the largest index of a non-zero component, i.e.

$$f(x) = \max\{k \leqslant n \mid x_i = 0,\ i = k+1, \dots, n\}.$$

This function is quasi-convex on $\mathbb{R}^n$. Its Lebesgue sets are subspaces of $\mathbb{R}^n$.

EXAMPLE 4.3.– Consider the function $f(x) = x_1x_2$ on $\mathbb{R}^2_+$. This function is not convex or concave since the matrix

$$f''(x) = \begin{pmatrix} 0 & 1 \\ 1 & 0 \end{pmatrix}$$

is indefinite. It has one positive and one negative eigennumber. But this function is quasi-concave, since all sets

$$\{x \in \mathbb{R}^2_+ \mid x_1x_2 \geqslant \beta\}$$

are convex.

EXAMPLE 4.4.– *(Fractional-linear functions)* The function

$$f(x) = \frac{\langle a, x\rangle + b}{\langle c, x\rangle + d}$$

with the domain of definition $\{x \mid \langle c, x\rangle + d > 0\}$ is quasi-convex (and quasi-concave), since all its Lebesgue sets

$$X_\beta = \left\{ x \,\middle|\, \langle c, x\rangle + d > 0,\ \frac{\langle a, x\rangle + b}{\langle c, x\rangle + d} \leqslant \beta \right\}$$

$$= \left\{ x \,\middle|\, \langle c, x\rangle + d > 0,\ \frac{\langle a - \beta c, x\rangle + b - \beta d}{\langle c, x\rangle + d} \leqslant 0 \right\}$$

are convex.

EXAMPLE 4.5.– *(Distance ratio)* Let $a, b \in \mathbb{R}^n$. Define the function

$$f(x) = \frac{\|x - a\|_2}{\|x - b\|_2},$$

that is, the ratio of distances from the point x to the points a and b. The function $f(x)$ is quasi-concave on $\{x \mid \|x - a\|_2 \leqslant \|x - b\|_2\}$. To show this, consider the Lebesgue set X_β for $\beta \leqslant 1$, since $f(x) \leqslant 1$ on $\{x \mid \|x - a\|_2 \leqslant \|x - b\|_2\}$. The Lebesgue set X_β is a set of points that satisfies condition

$$\|x - a\|_2 \leqslant \beta\|x - b\|_2.$$

We raise both sides of the inequality to a square. We get

$$(1 - \beta^2)\langle x, x\rangle - 2\langle a - \beta^2 b, x\rangle + \beta^2\langle b, b\rangle \leqslant 0.$$

This inequality describes a convex set (ball) if $\beta \leqslant 1$.

EXAMPLE 4.6.– *(Internal return rate)* Let the vector $x = (x_0, x_1, \ldots, x_n)$ determines the flow of payments for n periods, where $x_i > 0$ means that in the period i the firm receives payments, and $x_i < 0$ means that in the period i the firm pays. Determine the cost of the flow of payments in the following way:

$$\mathrm{PV}(x, r) = \sum_{i=0}^{n} (1+r)^{-i} x_i,$$

where r is the interest rate. The multiplier $(1 + r)^{-i}$ is a discount multiplier for payment in the period i.

Consider a flow of payments in which $x_0 < 0$ and $x_0 + x_1 + \ldots + x_n > 0$. This means that the firm first invests $|x_0|$ in the zero period and the sum of all other payments $x_1 + \ldots + x_n$ (not taking into account discounts multipliers) exceeds the initial investment.

For such a flow of payments $\mathrm{PV}(x, 0) > 0$ and $\mathrm{PV}(x, r) \to x_0 < 0$, when $r \to \infty$. It follows that at least for one $r \geqslant 0$ equality $PV(x, r) = 0$ holds true. Define the *internal return rate* as the lowest interest rate $r \geqslant 0$, for which the cost is zero:

$$\mathrm{IRR} = \inf\{r \geqslant 0 \,|\, \mathrm{PV}(x, r) = 0\}.$$

The internal rate of return is a quasi-concave function. To prove this, note that

$$\mathrm{IRR}(x) \geqslant R \Leftrightarrow \mathrm{PV}(x) \geqslant 0 \text{ for } 0 \leqslant r \leqslant R.$$

The expression on the left side determines the Lebesgue set of the function IRR. The expression on the right side is the intersection of sets $\{x \,|\, \mathrm{PV}(x, r) \geqslant 0\}$ with $0 \leqslant r \leqslant R$. For every r, the expression $\mathrm{PV}(x) \geqslant 0$ defines a subspace, hence, the expression on the right side determines a convex set.

The above examples show that the notion of quasi-convexity is an essential generalization of the notion of convexity. However, many properties of convex functions remain valid or have analogues for quasi-convex functions, for example, the analogue of Jensen's inequality, which characterizes quasi-convexity.

THEOREM 4.1.– A function $f : X \to \mathbb{R}$, where $X \subset \mathbb{R}^n$ is a convex set, is quasi-convex if and only if for all $x^1, x^2 \in X$, $0 \leqslant \lambda \leqslant 1$ the following inequality holds true:

$$f(\lambda x^1 + (1 - \lambda)x^2) \leqslant \max\{f(x^1), f(x^2)\}. \qquad [4.1]$$

PROOF.– Let a function f be quasi-convex, that is, the set X_β is convex for every β. We fix two arbitrary points $x^1, x^2 \in X$ and consider the point $x = \lambda x^1 + (1 - \lambda)x^2$,

$\lambda \in (0,1)$. Points x^1, $x^2 \in X_\beta$ for $\beta = \max\{f(x^1), f(x^2)\}$. Since the set X_β is convex, then $x \in X_\beta$, and $f(x) \leqslant \beta = \max\{f(x^1), f(x^2)\}$, that is, inequality [4.1] holds true.

Suppose now that the inequality [4.1] holds true. Fix two arbitrary points $x^1, x^2 \in X_\beta$. Then $\max\{f(x^1), f(x^2)\} \leqslant \beta$. Since the set X is convex, then for arbitrary $\lambda \in (0,1)$ the point $x = \lambda x^1 + (1-\lambda)x^2 \in X$. From inequality [4.1], it follows that $f(x) \leqslant \max\{f(x^1), f(x^2)\} \leqslant \beta$, that is $x \in X_\beta$. Hence the set X_β is convex and f is quasi-convex function. □

EXAMPLE 4.7.– *(The rank of a not-negative definite matrix)* The function $f(X) = \mathrm{Rank}(X)$ is quasi-concave on the set of all not-negative definite matrices of dimension $n \times n$. This follows from Jensen's inequalities for quasi-concave functions [4.1]

$$\mathrm{Rank}(X+Y) \geqslant \max\{\mathrm{Rank}(X), \mathrm{Rank}(Y)\},$$

where X, Y are a not-negative definite matrix of dimension $n \times n$.

We now give a simple characterization of quasi-convex functions on $\mathbb{R}$. We consider continuous functions because the formulation of the theorem in the general case is too complicated.

THEOREM 4.2.– A continuous function f: $X \to \mathbb{R}$, where X is a convex set in $\mathbb{R}$, is quasi-convex if and only if one of the following conditions hold true:

– function f is non-decreasing;

– function f is non-increasing;

– there exists a point $c \in X$, such that for all $t \in X$, $t \leqslant c$, the function f is non-increasing, and for all $t \in X$, $t \geqslant c$, the function f is non-decreasing.

4.1.1. *Differentiable quasi-convex functions*

THEOREM 4.3.– Let f: $X \to \mathbb{R}$ be a differentiable function on X, where $X \subset \mathbb{R}^n$ is an open convex set. The function f is quasi-convex on X if and only if

$$f(y) \leqslant f(x) \Rightarrow \langle f'(x), y - x\rangle \leqslant 0 \quad \text{for all} \quad x, y \in X. \qquad [4.2]$$

PROOF.– First we show that if the function f is quasi-convex on X, then relation [4.2] holds true. Consider arbitrary points $x, y \in X$ such that $f(y) \leqslant f(x)$. From the differentiability of the function $f(x)$ at point x for $\lambda \in (0,1)$, we have

$$f(\lambda y + (1-\lambda)x) - f(x) = \lambda\langle f'(x), y - x\rangle + \lambda\|y - x\|\alpha(x; \lambda(y - x)),$$

where $\alpha(x;\lambda(y-x)) \to 0$ as $\lambda \to 0$. Since the function f is quasi-convex, then $f(\lambda y+(1-\lambda)x) \leqslant f(x)$. Hence

$$\lambda\langle f'(x), y-x\rangle + \lambda\|y-x\|\alpha(x;\lambda(y-x)) \leqslant 0.$$

Dividing this inequality by λ and directing λ to zero, we get that $\langle f'(x), y-x\rangle \leqslant 0$.

Suppose now that relation [4.2] holds true. Consider arbitrary points x, $y \in X$, such that $f(y) \leqslant f(x)$. We have to prove that $f(\lambda y+(1-\lambda)x) \leqslant f(x)$ for all $\lambda \in (0,1)$. To do this, just show that the set

$$L = \{z \mid z = \lambda y + (1-\lambda)x,\ \lambda \in (0,1),\ f(z) > f(x)\}$$

is empty. Let it be wrong, that is, assume that there exists $x' \in L$. Then $x' = \lambda y + (1-\lambda)x$ for some $\lambda \in (0,1)$ and $f(x') > f(x)$. Since the function f is differentiable, then it is continuous and there exists $\delta \in (0,1)$ such that

$$f(\mu x' + (1-\mu)x) > f(x) \quad \text{for all} \quad \mu \in [\delta, 1],$$

and $f(x') > f(\delta x' + (1-\delta)x)$. From this inequality and the theorems about the mean value of the differentiable function, we get that

$$0 < f(x') - f(\delta x' + (1-\delta)x) = (1-\delta)\langle f'(\hat{x}), x'-x\rangle,$$

where $\hat{x} = \hat{\mu}x' + (1-\hat{\mu})x$ for some $\hat{\mu} \in (\delta, 1)$. It is obvious that $f(\hat{x}) > f(x)$. By dividing the previous inequality by $1-\delta > 0$, we get $\langle f'(\hat{x}), x'-x\rangle > 0$. It follows from this relation that

$$\langle f'(\hat{x}), y-x\rangle > 0.$$

From other side, $f(\hat{x}) > f(x) \geqslant f(y)$, and the point $\hat{x}$ is a convex combination of points x and y, $\hat{x} = \hat{\lambda}y + (1-\hat{\lambda})x$, $\hat{\lambda} \in (0,1)$. By assumption of the theorem, we have $\langle f'(\hat{x}), y-\hat{x}\rangle \leqslant 0$. Therefore, the relationship must be fulfilled

$$0 \geqslant \langle f'(\hat{x}), y-\hat{x}\rangle = (1-\hat{\lambda})\langle f'(\hat{x}), y-x\rangle.$$

The last inequality is incompatible with inequality $\langle f'(\hat{x}), y-x\rangle > 0$. Hence, $L = \varnothing$. □

Condition [4.2] has a simple geometric interpretation when $f'(x) \neq 0$. It claims that $f'(x)$ determines a support hyperplane to the Lebesgue set $\{y \mid f(y) \leqslant f(x)\}$.

COROLLARY 4.1.– Let $f\colon X \to \mathbb{R}$ be a two times differentiable function on X, where $X \subset \mathbb{R}^n$ is an open convex set. If the function f is quasi-convex on X, then the following condition holds true:

$$\langle y, f'(x)\rangle = 0 \Rightarrow \langle f''(x)y, y\rangle \geqslant 0 \quad \text{for all } x \in X,\ y \in \mathbb{R}^n. \qquad [4.3]$$

For quasi-convex functions on $\mathbb{R}$, this statement has the following form:

$$f'(x) = 0 \Rightarrow f''(x) \geqslant 0,$$

that is, at the point with zero slope the second derivative is non-negative. For quasi-convex functions on $\mathbb{R}^n$, interpretation of condition [4.3] is somewhat more complicated. As in one-dimensional case, if $f'(x) = 0$, then $f''(x) \geqslant 0$ must be fulfilled. If $f'(x) \neq 0$, then condition [4.3] means that $f''(x)$ is non-negative definite on the $(n-1)$-dimensional subspace $f'(x)^{\perp}$. This implies that the matrix $f''(x)$ must have at least one negative eigennumber. On the contrary, if f satisfies the condition

$$\langle y, f'(x)\rangle = 0 \Rightarrow \langle f''(x)y, y\rangle > 0$$

for all $x \in X$ and $y \in \mathbb{R}^n$, $y \neq 0$, then f is quasi-convex function. This condition means that the matrix $f''(x)$ is positive definite at each point x, where $f'(x) = 0$, and the matrix $f''(x)$ is positive definite on the $(n-1)$-dimensional subspace $f'(x)^{\perp}$ for other points.

4.1.2. *Operations that preserve quasi-convexity*

– The maximum of weighted quasi-convex functions with non-negative weights is a quasi-convex function, i.e.

$$f(x) = \max\{w_1 f_1(x), \dots, w_n f_n(x)\}$$

where $w_i > 0$ and f_i are quasi-convex functions.

– The supremum of a family of quasi-convex functions:

$$f(x) = \sup_{y \in C} w(y) g(x, y),$$

where $w(y) \geqslant 0$ and $g(x, y)$ is quasi-convex function with respect to x for all $y \in C$.

This statement can be easily verified: $f(x) \leqslant \beta$ if and only if

$$w(y)g(x,y) \leqslant \beta \quad \text{for all} \quad y \in C,$$

that is, the Lebesgue set X_β of the function $f(x)$ is the intersection of the Lebesgue sets $Z(y)_\beta$ of functions $w(y)g(x, y)$ with respect to x.

EXAMPLE 4.8.– *(Generalized eigenvalues)* The greatest generalized eigenvalue of pairs of matrices (X, Y), where $Y > 0$, is determined as follows:

$$\lambda_{\max}(X, Y) = \sup_{u \neq 0} \frac{\langle Xu, u\rangle}{\langle Yu, u\rangle} = \sup\{\lambda \mid \det(\lambda Y - X) = 0\}.$$

This function is quasi-convex on the set $\{(X, Y)\}$, where X is a symmetric $n \times n$ matrix and Y is a symmetric positive definite $n \times n$ matrix. To prove this, note that for each $u \neq 0$, the function $\langle Xu, u\rangle / \langle Yu, u\rangle$ is fractional-linear, and hence quasi-convex. So $\lambda_{\max}$ is the supremum of the family of quasi-convex functions.

– If $h\colon X \to \mathbb{R}$, where $X \subset \mathbb{R}^n$, is a quasi-convex function and $g\colon Y \to \mathbb{R}$, where $Y \subset \mathbb{R}$, is a non-decreasing function, then $f(x) = g(h(x))$ is a quasi-convex function.

– A quasi-convex function from an affine or fractional-linear function is quasi-convex.

– If g is a quasi-convex function, then $f(x) = g(Ax + b)$ is a quasi-convex function, and $f(x) = g((Ax + b)/(\langle c, x\rangle + d))$ is a quasi-convex function on the set

$$\left\{x \,\middle|\, \frac{Ax + b}{\langle c, x\rangle + d} \in Y, \langle c, x\rangle + d > 0\right\}.$$

THEOREM 4.4.– If $g(x, y)$ is a quasi-convex with respect to (x, y) function and C is a convex set, then the function

$$f(x) = \inf_{y \in C} g(x, y)$$

is quasi-convex.

PROOF.– By definition of the function f, $f(x) \leqslant \beta$ if and only if for all $\epsilon > 0$, there exists $y \in C$ such that $g(x, y) \leqslant \beta + \epsilon$. Let X_β be the Lebesgue set of the function f. Then for arbitrary $\epsilon > 0$, there exist $y_1, y_2 \in C$ such that

$$g(x_1, y_1) \leqslant \beta + \epsilon, \quad g(x_2, y_2) \leqslant \beta + \epsilon.$$

Since g is a quasi-convex with respect to (x, y) function, then

$$g(\theta x_1 + (1 - \theta)x_2, \theta y_1 + (1 - \theta)y_2) \leqslant \beta + \epsilon,$$

for $0 \leqslant \theta \leqslant 1$. Hence, $f(\theta x_1 + (1 - \theta)x_2) \leqslant \beta$. □

4.1.3. *Representation in the form of a family of convex functions.*

We find a family of convex functions $\phi_t\colon \mathbb{R}^n \to \mathbb{R}$ (the index t takes real values) such that

$$f(x) \leqslant t \Leftrightarrow \phi_t(x) \leqslant 0.$$

That is, if X_β, $\beta \geqslant 0$ is the Lebesgue set of the function f, and $Y_{t,\beta}$ is the Lebesgue set of the function ϕ_t, then $X_t = Y_{t,0}$. Obviously, the functions ϕ_t must satisfy the inequality $\phi_t(x) \leqslant 0 \Rightarrow \phi_s(x) \leqslant 0$ for $s \geqslant t$ and all $x \in \mathbb{R}^n$. This inequality holds if for any x the function $\phi_t(x)$ is non-increasing with respect to t.

To show that such a representation exists, let us take

$$\phi_t(x) = \begin{cases} 0, & f(x) \leqslant t; \\ \infty, & \text{in other cases.} \end{cases}$$

This presentation is not unique. For example, if Lebesgue sets are closed, then we can take

$$\phi_t(x) = \text{dist}\,(x, \{z \mid f(z) \leqslant t\}).$$

Of course, we are interested in the family ϕ_t with "good" properties, for example, differentiability.

EXAMPLE 4.9.– Let p be a convex function, and let q be a concave function, such that $p(x) \geqslant 0$ and $q(x) > 0$ on a convex set C. Then the function $f(x) = p(x)/q(x)$ is quasi-convex on C. For such a function, we have

$$f(x) \leqslant t \Leftrightarrow p(x) - tq(x) \leqslant 0,$$

and we can take the function ϕ_t of the form $\phi_t(x) = p(x) - tq(x)$ for $t \geqslant 0$. For each t, the function $\phi_t(x)$ is convex, and for each x, the function $\phi_t(x)$ is decreasing on t.

4.1.4. *The maximization problem for quasi-convex functions*

THEOREM 4.5.– Let X be a compact polyhedral set in $\mathbb{R}^n$, and let a function $f\colon \mathbb{R}^n \to \mathbb{R}$ be quasi-convex and continuous on X. Consider the optimization problem

$$f(x) \to \max, \quad x \in X.$$

Among the solutions to this problem there is necessarily an extreme point $\bar{x}$.

PROOF.– Since the function f is continuous, it reaches a maximum on X at some point $x' \in X$. If there is an extreme point in which the value of the objective function is $f(x')$, then the assertion is valid.

Suppose it is not, that is $f(x') > f(x^j)$, where $x^j, j = 1, \ldots, k$ are extreme points of the set X.

By theorem 2.26 (Minkowski theorem about a convex compact), the point x' can be represented in the form

$$x' = \sum_{j=1}^{k} \lambda_j x^j, \quad \sum_{j=1}^{k} \lambda_j = 1, \quad \lambda_j \geqslant 0, j = 1, \ldots, k,$$

where x^j, $j = 1, \ldots, k$ are extreme points of the set X. Since $f(x') > f(x^j)$ for all j, then

$$f(x') > \max_{1 \leqslant j \leqslant k} f(x^j) = \beta.$$

Consider the set $X_\beta = \{x \mid f(x) \leqslant \beta\}$. Note that $x^j \in X_\beta$ for $j = 1, \ldots, k$ and X_β is a convex Lebesgue set of the function $f(x)$. Hence, $x' = \sum_{j=1}^{k} \lambda_j x^j \in X_\beta$. Therefore $f(x') \leqslant \beta$. We came to a contradiction. This shows that $f(x') = f(x_j)$ for some extreme point x^j. □

4.1.5. *Strictly quasi-convex functions*

DEFINITION 4.2.– *Let X be a convex set in $\mathbb{R}^n$. A function $f\colon X \to \mathbb{R}$ is called* strictly quasi-convex, *if for all $x_1, x_2 \in X$, such that $f(x_1) \neq f(x_2)$, and all $\lambda \in (0,1)$ the inequality*

$$f(\lambda x_1 + (1-\lambda)x_2) < \max\{f(x_1), f(x_2)\}$$

holds true. A function f is called strictly quasi-concave *if the function $g = -f$ is strictly quasi-convex.*

It follows from the definition that any convex function is also strictly quasi-convex. By definition 3.1, a strictly convex function is convex. But a strictly quasi-convex function is not necessarily quasi-convex. Here is an example proposed by Karamardian:

$$f(x) = \begin{cases} 1, \text{ for } & x = 0, \\ 0, \text{ for } & x \neq 0. \end{cases}$$

By definition 4.2, the function $f(x)$ is strictly quasi-convex. But it is not quasi-convex since for $x_1 = 1$ and $x_2 = -1$, we have $f(x_1) = f(x_2) = 0$, and

$$f\left(\frac{1}{2}x_1 + \frac{1}{2}x_2\right) = f(0) = 1 > f(x_2).$$

If the function f is lower semicontinuous, then its strict quasi-convexity implies the usual quasi-convexity.

The following theorem shows that any local minimum of a strictly quasi-convex function on a convex set is also its global minimum. Quasi-convex functions do not have such a property.

THEOREM 4.6.– Let $f\colon \mathbb{R}^n \to \mathbb{R}$ be a strictly quasi-convex function. Consider the problem of minimization of $f(x)$ under the condition $x \in X$, where X is a convex set in $\mathbb{R}^n$. Let $\hat{x}$ be a point of local minimum of the problem. Then the point $\hat{x}$ is a point of global minimum of the problem.

PROOF.– Assume that the statement of the theorem is not true. Let there exist a point $\bar{x} \in X$, such that $f(\bar{x}) < f(\hat{x})$. It follows from the convexity of the set X that the point $\lambda\bar{x} + (1-\lambda)\hat{x} \in X$ for all $\lambda \in (0,1)$. Since $\hat{x}$ is a point of local minimum, then $f(\hat{x}) \leqslant f(\lambda\bar{x} + (1-\lambda)\hat{x})$ for all $\lambda \in (0,\delta)$ for some $\delta \in (0,1)$. Due to the strict quasi-convexity of the function f and inequality $f(\bar{x}) < f(\hat{x})$, we have that $f(\lambda\bar{x} + (1-\lambda)\hat{x}) < f(\hat{x})$ for all $\lambda \in (0,1)$. The obtained contradiction proves the theorem. □

LEMMA 4.1.– Let X be a convex set in $\mathbb{R}^n$, and let $f: X \to \mathbb{R}$ be a strictly quasi-convex lower semicontinuous function. Then the function f is quasi-convex.

PROOF.– Let $x_1, x_2 \in X$. If $f(x_1) \neq f(x_2)$, then according to the definition of the strict quasi-convexity of the function f for each $\lambda \in (0,1)$ we have

$$f(\lambda x_1 + (1-\lambda)x_2) < \max\{f(x_1), f(x_2)\}.$$

Let $f(x_1) = f(x_2)$. To make sure that the function f is quasi-convex, we need to show that

$$f(\lambda x_1 + (1-\lambda)x_2) < f(x_1)$$

for all $\lambda \in (0,1)$. Assume the opposite, that is, let $f(\mu x_1 + (1-\mu)x_2) > f(x_1)$ for some $\mu \in (0,1)$. Consider the point $x = \mu x_1 + (1-\mu)x_2$. Since the function f is lower semicontinuous, then there exists $\lambda \in (0,1)$, such that

$$f(x) > f(\lambda x_1 + (1-\lambda)x) > f(x_1) = f(x_2). \quad [4.4]$$

Note that the point x can be represented in the form of a convex combination of the points $\lambda x_1 + (1-\lambda)x$ and x_2. Then, since the function f is strictly quasi-convex and $f(\lambda x_1 + (1-\lambda)x) > f(x_2)$, we have $f(x) < f(\lambda x_1 + (1-\lambda)x)$. This contradicts [4.4]. □

4.1.6. *Strongly quasi-convex functions*

DEFINITION 4.3.– *Let* f: $X \to \mathbb{R}$, *where* X *is a convex set in* $\mathbb{R}^n$. *The function* f *is called* strongly quasi-convex, *if for all* $x_1, x_2 \in X$, $x_1 \neq x_2$, *and all* $\lambda \in (0,1)$ *the following inequality holds true:*

$$f(\lambda x_1 + (1-\lambda)x_2) < \max\{f(x_1), f(x_2)\}.$$

The function f *is called* strongly quasi-concave, *if the function* $g = -f$ *is strongly quasi-convex.*

The following statements follow the definitions:

– a strictly convex function is strongly quasi-convex;

– a strongly quasi-convex function is strictly quasi-convex;

– a strongly quasi-convex function is quasi-convex, even if it is not lower semicontinuous.

THEOREM 4.7.– Let f: $\mathbb{R}^n \to \mathbb{R}$ be a strongly quasi-convex function. Consider the problem of minimization of $f(x)$ under the condition $x \in X$, where X is a convex set in $\mathbb{R}^n$. Let $\hat{x}$ be a point of local minimum of the problem. Then the point $\hat{x}$ is a unique point of global minimum of the problem.

PROOF.– Since $\hat{x}$ is a point of local minimum of the problem, then there exists an ε neighborhood $N_\varepsilon(\hat{x})$ of the point $\hat{x}$, such that $f(\hat{x}) \leqslant f(x)$ for all $x \in X \cap N_\varepsilon(\hat{x})$. Assume that the statement of the theorem is not correct, that is, there is a point $\bar{x} \in X$, such that $\bar{x} \neq \hat{x}$ and $f(\bar{x}) < f(\hat{x})$. It follows from the strong quasi-convexity f that

$$f(\lambda\bar{x} + (1-\lambda)\hat{x}) < \max\{f(\bar{x}), f(\hat{x})\} = f(\hat{x})$$

for all $\lambda \in (0,1)$. But for sufficiently small λ

$$\lambda\bar{x} + (1-\lambda)\hat{x} \in X \cap N_\varepsilon(\hat{x}).$$

The last inequality contradicts the local optimality of $\hat{x}$ □

4.2. Pseudo-convex functions

DEFINITION 4.4.– *Let X be an open set in $\mathbb{R}^n$, and let f: $X \to \mathbb{R}$ be a differentiable function. The function f is called* pseudo-convex, *if for all $x_1, x_2 \in X$ such that*

$$\langle f'(x_1), x_2 - x_1 \rangle \geqslant 0,$$

the inequality $f(x_2) \geqslant f(x_1)$ holds true, or, which is equivalent, if $f(x_2) < f(x_1)$, then

$$\langle f'(x_1), x_2 - x_1 \rangle < 0.$$

The function f is called pseudo-concave, *if the function $g = -f$ is pseudo-convex.*

DEFINITION 4.5.– *The function f is called* strictly pseudo-convex, *if for all different $x_1, x_2 \in X$ such that*

$$\langle f'(x_1), x_2 - x_1 \rangle \geqslant 0,$$

the inequality $f(x_2) > f(x_1)$ *holds true, or, which is equivalent, if for all different* $x_1, x_2 \in X$ *from inequality* $f(x_2) \leqslant f(x_1)$ *follows the inequality*

$$\langle f'(x_1), x_2 - x_1 \rangle < 0.$$

The function f *is called* strictly pseudo-concave, *if the function* $g = -f$ *is strictly pseudo-convex.*

THEOREM 4.8.– Let X be an open set in $\mathbb{R}^n$, and let f: $X \to \mathbb{R}$ be a differentiable pseudo-convex function. Then the function f is strictly quasi-convex and quasi-convex.

PROOF.– Let us show first that the function f is strictly quasi-convex. Assume that this is not the case, that is, there are $x_1, x_2 \in X$, such that $f(x_1) \neq f(x_2)$ and $f(x') \geqslant \max\{f(x_1), f(x_2)\}$, where $x' = \lambda x_1 + (1-\lambda)x_2$ for some $\lambda \in (0,1)$. Let $f(x_1) < f(x_2)$. Consequently

$$f(x') \geqslant f(x_2) > f(x_1). \qquad [4.5]$$

It follows from the definition of the pseudo-convex function f that

$$\langle f'(x'),\ x_1 - x' \rangle < 0.$$

Since $\langle f'(x'), x_1 - x' \rangle < 0$ and

$$x_1 - x' = -\frac{1-\lambda}{\lambda}(x_2 - x'),$$

we have $\langle f'(x'), x_2 - x' \rangle > 0$. From the pseudo-convexity of the function f, we also have $f(x_2) \geqslant f(x')$. It follows from [4.5] that $f(x_2) = f(x')$. Since $\langle f'(x'), x_2 - x' \rangle > 0$, we can find a point $\hat{x} = \mu x' + (1-\mu)x_2$, $\mu \in (0,1)$, such that

$$f(\hat{x}) > f(x') = f(x_2).$$

Similarly, using the pseudo-convexity of f it is easy to make sure that $\langle f'(x), x_2 - \hat{x} \rangle < 0$ and $\langle f'(\hat{x}), x' - \hat{x} \rangle < 0$. Note that

$$x_2 - \hat{x} = \frac{\mu}{1-\mu}(\hat{x} - x').$$

Hence, the last two inequalities are incompatible. The resulting contradiction shows that the assumption was incorrect, that is, the function f is strictly quasi-convex. According to the previous lemma, it is also quasi-convex. □

THEOREM 4.9.– Let X be an open set in $\mathbb{R}^n$, and let f: $X \to \mathbb{R}$ be a differentiable strictly pseudo-convex function. Then the function f is strongly quasi-convex.

PROOF.– Assume that the statement of the theorem is incorrect, that is, there are different $x_1, x_2 \in X$ and $\lambda \in (0,1)$, such that

$$f(x) \geqslant \max\{f(x_1), f(x_2)\}, \quad x = \lambda x_1 + (1-\lambda)x_2.$$

Since $f(x_1) \leqslant f(x)$, then it follows from the strict pseudo-convexity of the function f that $\langle f'(x), x_1 - x\rangle < 0$. Hence

$$\langle f'(x), x_1 - x_2\rangle < 0.$$

Also, since $f(x_2) \leqslant f(x)$, then

$$\langle f'(x), x_2 - x_1\rangle < 0.$$

The last two inequalities contradict each other. Hence, the function f is strongly quasi-convex. □

EXAMPLE 4.10.– The function

$$f_1(x) = \begin{cases} 0, & \text{for} \quad x \in \mathbb{R}\backslash\mathbb{R}_+; \\ -x^n, & \text{for} \quad x \in \mathbb{R}_+,\ n \in \mathbb{N},\ n \geq 2, \end{cases}$$

is quasi-convex on the set $\mathbb{R}$. However, it is not strictly quasi-convex, not pseudo-convex, not strictly pseudo-convex, not convex.

EXAMPLE 4.11.– The function

$$f_2(x) = -x^n,\ x \in \mathbb{R}_+,\ n \in \mathbb{N},\ n \geq 2,$$

is quasi-convex and strictly quasi-convex on the set $\mathbb{R}_+$. However, it is not pseudo-convex, strictly pseudo-convex, nor convex.

EXAMPLE 4.12.– The function

$$f_3(x) = -(x_1+1)^2 + 1, \quad x = (x_1, x_2) \in \mathbb{R}^2_+,$$

is pseudo-convex, strictly quasi-convex and quasi-convex on the set $\mathbb{R}^2_+$. However, it is not strictly pseudo-convex, nor convex.

EXAMPLE 4.13.– The function

$$f_4(x) = ax + b, \quad x \in \mathbb{R}, \quad a < 0,$$

is convex, strictly pseudo-convex, pseudo-convex, strictly quasi-convex and quasi-convex on the set $\mathbb{R}$. However, it is not strictly convex.

EXAMPLE 4.14.– The function

$$f_5(x) = x^n, \quad x \in \mathbb{R}_+,\ n \in \mathbb{N},\ n \geq 2,$$

is strictly convex, convex, strictly pseudo-convex, pseudo-convex, strictly quasi-convex and quasi-convex on the set $\mathbb{R}_+$.

4.3. Logarithmically convex functions

DEFINITION 4.6.– *A function* f: $X \to \mathbb{R}$, $X \subset \mathbb{R}^n$ *is called* logarithmically convex, *if* $f(x) > 0$ *for all* $x \in X$ *and the function* $\ln f(x)$ *is convex. A function* f *is called* logarithmically concave, *if the function* $g = -\ln f(x)$ *is convex. Thus, a function* f *is logarithmically concave if and only if the function* $1/f$ *is logarithmically convex.*

Since the function $g = e^f$ is convex when the function f is convex it follows that the logarithmically convex function is convex. Similarly, the concave function is logarithmically concave. In addition, the logarithmically convex function is quasi-convex, and the logarithmically concave function is quasi-concave.

EXAMPLE 4.15.– Let us consider a few simple examples:

– the affine function $f(x) = \langle a, x\rangle + b$ is logarithmically concave on $\{x \mid \langle a, x\rangle + b > 0\}$;

– the function $f(x) = x^a$ on $\operatorname{int} \mathbb{R}_+$ is logarithmically convex for $a \leqslant 0$ and it is logarithmically concave for $a \geqslant 0$;

– the exponential function $f(x) = e^{ax}$ is logarithmically convex and logarithmically concave;

– the function of normal distribution $F(x) = \frac{1}{\sqrt{2\pi}} \int_{-\infty}^{x} e^{-u^2/2} du$ is logarithmically concave;

– the Gamma function $\Gamma(x) = \int_0^\infty u^{x-1} e^{-u} du$ is logarithmically convex for $x \geqslant 1$;

– determinant $\det X$ is logarithmically convex on the set of all positive definite matrices S^n_{++}.

EXAMPLE 4.16.– *(The logarithmically concave densities of distributions)* Many probabilistic distributions have logarithmically concave densities. For example, the density of a multidimensional normal distribution

$$f(x) = \frac{1}{\sqrt{(2\pi)^n \det \Sigma}} e^{-\frac{1}{2}(x-\bar{x})^T \Sigma^{-1}(x-\bar{x})},$$

where $\bar{x} \in \mathbb{R}^n$, Σ is a positive definite matrix of dimension $n \times n$, $\Sigma \in S^n_{++}$, the density of exponential distribution on $\mathbb{R}^n_+$

$$f(x) = \left(\prod_{i=1}^n \lambda_i\right) e^{-\langle \lambda, x\rangle}, \ \lambda \geqslant 0.$$

Another example is the density of a uniform distribution on a convex set C:

$$f(x) = \begin{cases} \frac{1}{\alpha}, & x \in C; \\ 0, & x \notin C, \end{cases}$$

where α is the Lebesgue measure of the set C ($\ln f(x) = -\infty, x \notin C$).

A more exotic example is the Wishart distribution density, which is determined as follows. Let $x_1, \dots, x_p \in \mathbb{R}^n, p > n$, be independent normally distributed random vectors with zero mean and covariance matrix Σ. The random matrix $X = \sum_{i=1}^{p} x_i x_i^\top$ has the Wishart distribution with the density

$$f(X) = a(\det X)^{(p-n-1)/2} e^{-\frac{1}{2}\operatorname{Tr}\Sigma^{-1}X},$$

where X is a positive definite matrix and a is a positive constant. The Wishart distribution density is logarithmically concave, since the function

$$\ln f(X) = \ln a + \frac{p-n-1}{2} \ln \det X - \frac{1}{2} \operatorname{Tr}\Sigma^{-1}X$$

is concave with respect to X.

4.3.1. *Properties of logarithmically convex functions*

THEOREM 4.10.– Let f: $X \to \mathbb{R}$ be a two times differentiable function on a convex set X. The function f is logarithmically convex if and only if

$$f(x)f''(x) \geqslant f'(x)(f'(x))^\top \quad \text{for all} \quad x \in X,$$

and it is logarithmically concave if and only if

$$f(x)f''(x) \leqslant f'(x)(f'(x))^\top \quad \text{for all} \quad x \in X.$$

Obviously, the product and the scale change preserve the logarithmic convexity and logarithmic concavity.

Simple examples show that the sum of logarithmically concave functions may not be a logarithmically concave function. But the sum of logarithmically convex functions is a logarithmically convex function.

Let f and g be logarithmically convex functions, that is $F = \ln f$ and $G = \ln g$ are convex functions. By the theorem on a complex function

$$\ln(\exp F + \exp G) = \ln(f + g)$$

is a convex function. Consequently, the sum of two logarithmically convex functions is logarithmic convex function.

In general, if the function $f(x, y)$ is logarithmically convex with respect to x for all $y \in C$, then the function

$$g(x) = \int_C f(x, y)\, dy$$

is logarithmically convex.

EXAMPLE 4.17.– *(The Laplace transform of a non-negative function, the moment generating function and the cumulant generating function)* Let a function p: $\mathbb{R}^n \to \mathbb{R}$, and let $p(x) \geqslant 0$ for all x. The Laplace transform of the function p

$$P(z) = \int p(x) e^{-\langle z, x \rangle} dx,$$

is logarithmically convex. Let $p(x)$ be the density of a distribution function, that is $\int p(x)dx = 1$. The function $M(z) = P(-z)$ is called *the moment generating function* of the density $p(x)$. It has this name because the moments of the random variable ξ with density $p(x)$ are values of the derivatives of the moment generating function at point 0, that is

$$M'(0) = \mathbf{E}\xi, \quad M''(0) = \mathbf{E}\xi\xi^\top.$$

The function $\ln M(z)$, which is convex, is called *the cumulant generating function* of the density $p(x)$, since its derivatives yield cumulants of density. For example, the first and the second derivatives of the cumulant generating function at zero are, respectively, the mean and the covariation matrix of the random variable ξ:

$$(\ln M(x))'\big|_{x=0} = \mathbf{E}\xi, \quad (\ln M(x))''\big|_{x=0} = \mathbf{E}(\xi - \mathbf{E}\xi)(\xi - \mathbf{E}\xi)^\top.$$

4.3.2. *Integration of logarithmically concave functions*

THEOREM 4.11.– Let f: $\mathbb{R}^n \times \mathbb{R}^m \to \mathbb{R}$ be a logarithmically concave function with respect to x for all y. Then the function

$$g(x) = \int f(x, y) dy \qquad [4.6]$$

is logarithmically concave on $\mathbb{R}^n$.

This theorem has many important consequences. For example, it follows from this theorem that the marginal distributions of random variables with logarithmically concave densities are logarithmically concave. A few more consequences:

COROLLARY 4.2.– Let the functions f and g be logarithmically concave on $\mathbb{R}^n$. Then the convolution of functions f and g

$$(f * g)(x) = \int f(x - y) g(y) dy$$

is logarithmically concave on $\mathbb{R}^n$.

COROLLARY 4.3.– Let $C \subset \mathbb{R}^n$ be a convex set, and let ξ be a random vector from $\mathbb{R}^n$ with logarithmically concave density $p(x)$. Then the function

$$f(x) = \mathbf{P}(x + \xi \in C)$$

is logarithmically concave on x.

PROOF.– To show this, we write f in the form

$$f(x) = \int g(x+z)p(z)dz,$$

where g is determined in the following way:

$$g(u) = \begin{cases} 1, & u \in C; \\ 0, & u \notin C \end{cases}$$

(g is logarithmically concave). It remains to be used in [4.6]. □

EXAMPLE 4.18.– *(The distribution function of a random variable)* Let ξ be a random variable with the distribution function $F(x)$, that is

$$F(x) = \mathbf{P}(\xi < x).$$

If the density $f(x)$ of the distribution function $F(x)$ is logarithmically concave, then the distribution function $F(x)$ is also logarithmically concave.

EXAMPLE 4.19.– *(Volume of n-dimensional polyhedron)* Let A be a matrix of dimension $m \times n$. Determine the polyhedron

$$\mathcal{P}_u = \{x \in \mathbb{R}^n \mid Ax \leqslant u\}.$$

The volume $\mathrm{vol}\,\mathcal{P}_u$ of the polyhedron is a logarithmically concave function on u. To prove this, note that the function

$$\Psi(x,u) = \begin{cases} 1, & Ax \leqslant u, \\ 0, & \text{in other cases} \end{cases}$$

is logarithmically concave. From [4.6], it follows that the volume of the polyhedron

$$\mathrm{vol}\,\mathcal{P}_u = \int \Psi(x,u)\,du$$

is a logarithmically concave function.

4.4. Convexity in relation to order

DEFINITION 4.7.– *Let $K \subset \mathbb{R}^n$ be a right cone that determines the order relation $\preceq_K$ on $K \subset \mathbb{R}^n$. A function f: $\mathbb{R}^n \to \mathbb{R}$ is called K-*non-decreasing*, if*

$$x \preceq_K y \Rightarrow f(x) \leqslant f(y),$$

*and K-*increasing*, if*

$$x \preceq_K y,\ x \neq y \Rightarrow f(x) < f(y).$$

The K-non-increasing *and* K-decreasing *functions are determined in a similar way.*

EXAMPLE 4.20.– A function f: $\mathbb{R}^n \to \mathbb{R}$ is non-decreasing relative to $\mathbb{R}^n_+$, if

$$x_1 \leqslant y_1, \ldots, x_n \leqslant y_n \Rightarrow f(x) \leqslant f(y)$$

for all x, y. It is the same when f is non-decreasing with respect to any of its components.

EXAMPLE 4.21.– *(Matrix monotone functions)* A function $f\colon \mathbb{S}^n \to \mathbb{R}$ is called a *matrix monotone* (increasing and decreasing), if it is monotone (increasing and decreasing) relative to the cone of non-negative definite matrices.

Here are some examples of matrix monotone functions from $X \in \mathbb{S}^n$:

– the function $\mathrm{Tr}(WX)$, where $W \in \mathbb{S}^n$, is matrix non-decreasing if W is non-negative definite, and it is matrix increasing if W is positive definite (it is matrix non-increasing if W is non-positive definite, and it is matrix decreasing if W is negative definite);

– the function $\mathrm{Tr}\, X^{-1}$ is matrix decreasing on the set of positive definite matrices;

– the function $\det X$ is matrix increasing on the set of non-negative definite matrices.

THEOREM 4.12.– A differentiable function $f\colon X \to \mathbb{R}$ is K-non-decreasing if and only if

$$f'(x) \succeq_{K^*} 0 \quad \text{for all} \quad x \in X. \qquad [4.7]$$

The inverse statement is incorrect.

We will point out the difference from the scalar case: the derivative must be non-negative according to the conjugate order relation. In the case of strict monotonicity if

$$f'(x) \succ_{K^*} 0 \quad \text{for all} \quad x \in X, \qquad [4.8]$$

then f is K-increasing. The inverse statement is not correct.

PROOF.– First, let us assume that [4.7] holds true for all x, but f is not K-non-decreasing, that is there exist x and y, such that $x \preceq_{K^*} y$ and $f(y) < f(x)$. From the differentiability of the function f, it follows that there exists $t \in [0, 1]$, such that

$$\frac{d}{dt} f(x + t(y - x)) = \langle f'(x + t(y - x)), y - x \rangle < 0.$$

Since $y - x \in K$, this means that

$$-f'(x + t(y - x)) \notin K^*.$$

This contradicts the assumption that [4.7] holds true for all $x \in X$. We can also prove that from [4.8] K-increasing of the function f follows.

We can also show that condition [4.7] is necessary for K-non-decreasing of the function f. Suppose that condition [4.8] is not satisfied for $x = z$. Then, by the definition of a conjugate cone, there exists a $v \in K$ such that

$$\langle f'(z), v\rangle < 0.$$

Now consider $h(t) = f(z+tv)$ as a function from t. Since $h'(0) = \langle f'(z), v\rangle < 0$, then there exists $t > 0$, such that $h(t) = f(z + tv) < h(0) = f(z)$. This means that the function f is not K-non-decreasing. □

DEFINITION 4.8.– *Let $K \subset \mathbb{R}^m$ be a right cone that determines the order relation $\preceq_K$. A function f: $\mathbb{R}^n \to \mathbb{R}^m$ is called* K-convex, *if for all x, y and $0 \leqslant \lambda \leqslant 1$ the inequality*

$$f(\lambda x + (1 - \lambda)y) \preceq_K \lambda f(x) + (1 - \lambda)f(y)$$

holds true. A function f is called strictly K-convex, *if*

$$f(\lambda x + (1 - \lambda)y) \prec_K \lambda f(x) + (1 - \lambda)f(y).$$

for all x, y and $0 < \lambda < 1$. These definitions turn into definitions of convexity and strict convexity in the case where $m = 1$ and $K = \mathbb{R}_+$.

EXAMPLE 4.22.– A function f: $\mathbb{R}^n \to \mathbb{R}^m$ is convex componentwise, if for all x, y and $0 \leqslant \lambda \leqslant 1$

$$f(\lambda x + (1 - \lambda)y) \leqslant \lambda f(x) + (1 - \lambda)f(y),$$

that is, when each component $f_i, i = 1, \ldots, m$, is a convex function. The function f is strictly convex componentwise, if each of its components is strictly convex. In this case, the order relation is determined by the cone $\mathbb{R}^m_+$.

EXAMPLE 4.23.– *(Matrix convexity)* Let the function f take values on the set of symmetric matrices, f: $\mathbb{R}^n \to \mathbb{S}^m$. The function f is convex in accordance with the matrix order relation (that is, in accordance with the cone of non-negative definite matrices), if

$$f(\lambda x + (1 - \lambda)y) \leqslant \lambda f(x) + (1 - \lambda)f(y)$$

for all x, y and $\lambda \in [0,1]$. This is sometimes called *matrix convexity*. Equivalent definition: the function $\langle f(x)z, z\rangle$ is convex for all $z \in \mathbb{R}^m$.

This definition indicates the method of checking the matrix convexity.

A function f is strictly matrix convex, if

$$f(\lambda x + (1-\lambda)y) < \lambda f(x) + (1-\lambda)f(y)$$

for all $x \neq y$ and $0 < \lambda < 1$ or, equivalently, if the function $\langle f(x)z, z\rangle$ is strictly convex $z \neq 0$.

Here are some examples:

– the function $f(X) = X \cdot X^\top$, where X is a matrix of dimension $n \times m$, is convex, since for a fixed z the function $\langle XX^\top z, z\rangle = \|X^\top z\|^2$ is a convex quadratic form of component of the matrix X. For the same reason, the function $f(X) = X^2$ is convex on $\mathbb{S}^n$;

– the function X^p is matrix convex on a set of positive definite matrices, if $1 \leqslant p \leqslant 2$ or $-1 \leqslant p \leqslant 0$. The function X^p is matrix concave if $0 \leqslant p \leqslant 1$;

– the function $f(X) = e^X$ is not convex on the set of symmetric functions.

THEOREM 4.13.– A function f is K-convex if and only if for each $w \succeq_K 0$ the function $\langle f, w\rangle$ is convex. A function f is strictly K-convex if and only if for each non-zero $w \succeq_K 0$, the function $\langle f, w\rangle$ is strictly convex.

PROOF.– The assertion follows from the definition and properties of the conjugate order relation. □

THEOREM 4.14.– A differentiable function f: $X \to \mathbb{R}^m$ is K-convex if and only if for all x, $y \in X$

$$f(y) \succeq_K f(x) + \frac{\partial f(x)}{\partial x}(y-x).$$

A differentiable function f is strictly K-convex if and only if for all x, $y \in X$, $x \neq y$

$$f(y) \succ_K f(x) + \frac{\partial f(x)}{\partial x}(y-x).$$

THEOREM 4.15.– Let g: $X \to \mathbb{R}^m$, $X \subset \mathbb{R}^n$, be a K-convex function, let h: $U \to \mathbb{R}$ be a convex K-non-decreasing function on a convex set $U \subset \mathbb{R}^m$ and let $g(X) \subset U$. Then the function $h(g(x))$ is convex on X.

PROOF.– The function g is a K-convex function. Hence,

$$g(\lambda x_1 + (1-\lambda x_2)) \preceq_K \lambda g(x_1) + (1-\lambda)g(x_2).$$

The function h is K-non-decreasing and a convex function. Hence,

$$h(g(\lambda x_1 + (1-\lambda)x_2)) \leqslant h(\lambda g(x_1) + (1-\lambda)g(x_2)))$$
$$\leqslant \lambda h(g(x_1)) + (1-\lambda)h(g(x_2))).$$

□

EXAMPLE 4.24.– The quadratic form of the matrix

$$g(X) = X^\top AX + B^\top X + X^\top B + C,$$

where $A \in \mathbb{S}^m$, B is a matrix of dimension $m \times n$ and $C \in \mathbb{S}^n$, is convex if A is a positive definite matrix.

The function h: $\mathbb{S}^n \to \mathbb{R}$, $h(Y) = -\ln\det(-Y)$ is convex and increasing on the set of all symmetric negative definite matrices. By the preceding theorem, the function

$$f(X) = -\ln\det\left(-(X^\top AX + B^\top X + X^\top B + C)\right)$$

is convex on the set

$$\{X \mid X^\top AX + B^\top X + X^\top B + C < 0\}.$$

4.5. Exercises

1) For each of the following functions determine whether it is convex, concave, quasiconvex, or quasiconcave.

a) $f(x) = e^x - 1$ on $\mathbb{R}$.

b) $f(x_1, x_2) = -x_1x_2$ on $\mathbb{R}^2_{++}$.

c) $f(x_1, x_2) = 1/(x_1x_2)$ on $\mathbb{R}^2_{++}$.

d) $f(x_1, x_2) = x_1/x_2$ on $\mathbb{R}^2_{++}$.

e) $f(x_1, x_2) = x_1^2/x_2$ on $\mathbb{R} \times \mathbb{R}_{++}$.

f) $f(x_1, x_2) = x_1^\alpha x_2^{1-\alpha}$, where $0 \leq \alpha \leq 1$, on $\mathbb{R}^2_{++}$.

2) Consider a quadratic function f: $\mathbb{R}^n \to \mathbb{R}$ of the form $f(x) = \langle Hx, x\rangle$. The function f is called *positive subdefinite*, if from inequality $\langle Hx, x\rangle < 0$ follows that $Hx > 0$, or $Hx < 0$ for arbitrary $x \in \mathbb{R}^n$. Prove that the function f is quasi-convex on $\mathbb{R}^n_+$ if and only if it is positive subdefinite.

3) Consider a quadratic function f: $\mathbb{R}^n \to \mathbb{R}$ of the form $f(x) = \langle Hx, x\rangle$. The function f is called *strictly positive subdefinite*, if from inequality $\langle Hx, x\rangle < 0$, it follows that $Hx \geqslant 0$, or $Hx \leqslant 0$ for arbitrary $x \in \mathbb{R}^n$. Prove that the function f is pseudo-convex on $\mathbb{R}^n_+/\{0\}$ if and only if it is strictly positive subdefinite.

4) Let the following functions be given: $f_0, \ldots, f_n : \mathbb{R}_+ \to \mathbb{R}$. Consider the problem of approximation of the function f_0 by linear combinations of the functions $f_1, \ldots, f_n$. For $x \in \mathbb{R}^n$, the function $f = x_1 f_1 + \ldots x_n f_n$ approximates the function f_0 with accuracy $\varepsilon > 0$ on the interval $[0, T]$, if $|f(t) - f(t_0)| \leqslant \varepsilon$ for $0 \leqslant t \leqslant T$. Now fix the accuracy $\varepsilon > 0$ and define the *approximation length* as the largest T for which the function $f = x_1 f_1 + \ldots x_n f_n$ approximates f_0 with accuracy $\varepsilon > 0$ on the interval $[0, T]$:

$$W(x) = \sup\{T : |x_1 f_1(t) + \ldots + x_n f_n(t) - f_0(t)| \leqslant \varepsilon, 0 \leqslant t \leqslant T\}.$$

Show that the function W is quasi-concave.

5) A quasi-linear function on $\mathbb{R}$ (quasi-convex and quasi-concave) is monotone. Consider a generalization of this property to the case of functions on $\mathbb{R}^n$. Let the function f: $\mathbb{R}^n \to \mathbb{R}$ be quasi-linear. We consider it to be continuous. Show that it can be represented in the form $f(x) = g(\langle a, x\rangle)$, where g: $\mathbb{R} \to \mathbb{R}$ is a monotone function, and $a \in \mathbb{R}^n$. In other words, the quasi-linear function is a monotone function from the linear function. The inverse is also true.

6) Let c_1, c_2 be non-zero vectors from $\mathbb{R}^n$, let $\alpha_1, \alpha_2 \in \mathbb{R}$ and let $X = \{x \in \mathbb{R}^n \mid \langle c_2, x\rangle + \alpha_2 > 0\}$. The function f: $X \to \mathbb{R}$ is determined by the ratio

$$f(x) = \frac{\langle c_1, x\rangle + \alpha_1}{\langle c_2, x\rangle + \alpha_2}.$$

Show that the function f is pseudo-linear (both pseudo-convex and pseudo-concave).

7) Let g: $X \to \mathbb{R}$, h: $X \to \mathbb{R}$, where X is a convex set in $\mathbb{R}^n$. Show that the function f: $X \to \mathbb{R}$ of the form $f(x) = g(x)/h(x)$ is quasi-convex if the following conditions hold:

a) g is a convex function on X and $g(x) \geqslant 0$ for all $x \in X$;

b) h is a concave function on X and $h(x) > 0$ for all $x \in X$.

8) Let g: $X \to \mathbb{R}$, h: $X \to \mathbb{R}$, where X is a convex set in $\mathbb{R}^n$. Show that the function f: $X \to \mathbb{R}$, of the form $f(x) = g(x)/h(x)$ is quasi-convex if the following conditions hold:

a) g is a convex function on X and $g(x) \leqslant 0$ for all $x \in X$;

b) h is a convex function on X and $h(x) > 0$ for all $x \in X$.

9) Let g: $X \to \mathbb{R}$, h: $X \to \mathbb{R}$, where X is a convex set in $\mathbb{R}^n$. Show that the function f: $X \to \mathbb{R}$, of the form $f(x) = g(x)h(x)$ is quasi-convex if the following conditions hold:

a) g is a convex function on X and $g(x) \leqslant 0$ for all $x \in X$;

b) h is a concave function on X and $h(x) > 0$ for all $x \in X$.

10) Show that the functions defined in exercises 7–9 are pseudo-convex if X is an open set, and g, h are differentiable functions.

11) Let $f: \mathbb{R}^n \to \mathbb{R}^m$, $g: \mathbb{R}^n \to \mathbb{R}^k$ be differentiable and convex functions, and let the function $\phi: \mathbb{R}^{m+k} \to \mathbb{R}$ have the following property: if $a_2 \geqslant a_1$ and $b_2 \geqslant b_1$, then $\phi(a_2, b_2) \geqslant \phi(a_1, b_1)$. Consider the function $h: \mathbb{R}^n \to \mathbb{R}$ of the form $h(x) = \phi(f(x), g(x))$. Show that:

a) if ϕ is a convex function, then h is a convex function;

b) if ϕ is pseudo-convex function, then h is pseudo-convex function;

c) if ϕ is quasi-convex function, then h is quasi-convex function.

12) Show that the function $f(x) = e^x/(1+e^x)$, which is called a *logistic function*, is logarithmically concave.

13) Prove that the harmonic average

$$H(x) = \frac{1}{\dfrac{1}{x_1} + \ldots + \dfrac{1}{x_n}}$$

of $x_1, \ldots, x_n > 0$ is logarithmically concave.

14) Prove that if f: $X \to \mathbb{R}$ is logarithmically concave and $a \geqslant 0$, then the function $g = f - a$ is logarithmically concave on the set $\{x \in X \mid f(x) > a\}$.

15) Let P be a polynomial of $x \in \mathbb{R}$ with real roots. Prove that it is logarithmically concave on any interval where it is positive.

16) Let y be a random vector from $\mathbb{R}^n$ with logarithmically concave density, and let $g_i(x, y)$, $i = 1, \ldots, r$ be concave functions on $\mathbb{R}^m \times \mathbb{R}^n$. Then

$$h(x) = \mathbf{P}\left(g_1(x, y) \geqslant 0, \ldots, g_n(x, y) \geqslant 0\right)$$

is logarithmically concave on x. A special case is

$$h(x) = \mathbf{P}\left(g_1(x) \geqslant y_1, \ldots, g_r(x) \geqslant y_r\right),$$

where $g_i(x)$ are concave functions and y_i have logarithmically concave densities.

17) Let f: $\mathbb{R}_+ \to \mathbb{R}$ be non-negative functions. For $x \geqslant 0$, define

$$M(x) = \int_0^\infty u^x f(u) du.$$

When x is a positive number and f is a density of distribution, then $M(x)$ is the xth moment of a random variable with the density f. Show that M is a logarithmically concave function.

Hint: for each $u \geqslant 0$, the function u^x is logarithmically convex on $\operatorname{int} \mathbb{R}_+$.

Use the proved statement to show that the Gamma function

$$\Gamma(x) = \int_0^\infty u^{x-1}e^{-u}du$$

is logarithmically convex on $x \geqslant 1$.

18) The normal distribution function

$$f(x) = \frac{1}{\sqrt{2\pi}}\int_{-\infty}^{x} e^{-t^2/2}dt$$

is a logarithmically concave function. This follows from the general result that the convolution of two logarithmically concave functions is logarithmically concave. We give a simple proof that f is logarithmically concave function without reference to this result.

We note that f is logarithmically concave if and only if $f''(x)f(x) \leqslant (f'(x))^2$ for all x.

a) Check that f is logarithmically concave if $x \geqslant 0$.

b) Check that for all t and x the inequality $t^2/2 \geqslant -x^2/2 + xt$ holds true.

c) Using (b) show that $e^{-t^2/2} \leqslant e^{x^2/2-xt}$. Making use of this prove that

$$\int_{-\infty}^{x} e^{-t^2/2}dt \leqslant e^{x^2/2}\int_{-\infty}^{x} e^{-xt}dt.$$

d) Use (c) for checking the inequality $f''(x)f(x) \leqslant (f'(x))^2$ for $x \leqslant 0$.

19) Let $g(t) = \exp(-h(t))$ be a differentiable logarithmically concave density of distribution and let

$$f(x) = \int_{-\infty}^{x} g(t)dt = \int_{-\infty}^{x} e^{-h(t)}dt$$

be its distribution function. We show that f is logarithmically convex, i.e. it satisfies the relationship $f''(x)f(x) \leqslant (f'(x))^2$.

a) Express derivatives of f through the function h and its derivatives. Check that f is logarithmically concave if $h'(x) \geqslant 0$.

b) Assume that $h'(x) < 0$. Use inequality

$$h(t) \geqslant h(x) + h'(x)(t-x)$$

(which follows from the convexity of h) to prove that

$$\int_{-\infty}^{x} e^{-h(t)}dt \leqslant \frac{e^{-h(x)}}{-h'(x)}.$$

Use this inequality to check the logarithmic concavity of f.

20) The probability measure π on $\mathbb{R}^n$ is logarithmically concave if

$$\pi((1-\lambda)C_1 + \lambda C_2) \geqslant \pi(C_1)^{1-\lambda}\pi(C_2)^\lambda$$

for all convex subsets C_1 and C_2 from $\mathbb{R}^n$ and all $\lambda \in [0,1]$.

Show that when the measure π is generated by the density p, that is, $\pi(A) = \int_A p(x)dx$, then it is logarithmically concave if and only if the density p is logarithmically concave.

21) Show that the following densities are logarithmically concave:

a) The density of Gamma distribution

$$f(x) = \frac{\alpha^\lambda}{\Gamma(\lambda)} x^{\lambda-1} e^{-\alpha x}, \quad x \geqslant 0,$$

where $\lambda \geqslant 1, \alpha > 0$.

b) The density of a multidimensional hyperbolic distribution

$$f(x) = ce^{-\alpha\left(\delta + \langle \Sigma^{-1}(x-\bar{x}),(x-\bar{x})\rangle\right)^{1/2} + \langle \beta, x-\bar{x}\rangle},$$

where Σ is a positive definite symmetric matrix, $\beta \in \mathbb{R}^n$ and α and c are positive constants.

c) The density of Dirichlet distribution

$$f(x) = \frac{\Gamma(\lambda)}{\Gamma(\lambda_1)\cdots\Gamma(\lambda_{n+1})} x_1^{\lambda_1-1}\cdots x_n^{\lambda_n-1}\left(1-\sum_{i=1}^n x_i\right)^{\lambda_{n+1}-1}$$

on the set $\{x \in \mathbb{R}_+^n \mid \sum_{i=1}^n x_i \leqslant 1\}$. Here $\lambda_i > 0, i = 1,\ldots,n+1$, and $\lambda = \sum_{i=1}^{n+1} \lambda_i > 1$.

22) Show that the function

$$f(x) - \frac{\prod_{i=1}^n x_i}{\sum_{i=1}^n x_i}$$

is logarithmically concave on $\operatorname{int}\mathbb{R}_+^n$.

23) Show that the function $f(X) = X^{-1}$ is matrix convex on the cone of positive definite matrices.

24) Let $K \subset \mathbb{R}^m$ be a convex cone that determines the order relation. Show that two times differentiable function f: $X \to \mathbb{R}^m$, where $X \subset \mathbb{R}^n$, is K-convex if for all $x \in X$ and $y \in \mathbb{R}^n$ the inequality

$$\sum_{i,j=1}^n \frac{\partial^2 f}{\partial x_i \partial x_j} y_i y_j \succeq_K 0,$$

holds true, that is, the second derivative of the function f is a K-non-negative bilinear form. Here $\frac{\partial^2 f}{\partial x_i \partial x_j} \in \mathbb{R}^m$, with the components $\frac{\partial^2 f_k}{\partial x_i \partial x_j}$, $k = 1, \ldots, m$.

25) Let $K \subset \mathbb{R}^m$ determine a generalized order relation and let f: $X \to \mathbb{R}^m$, where $X \subset \mathbb{R}^n$. The Lebesgue sets of the function f (in accordance to the generalized order relation $\preceq_K$) are determined in the following way:

$$X_\beta = \{x \in X \mid f(x) \preceq_K \beta\},$$

where $\beta \in \mathbb{R}^m$.

Epigraph of the function f relative to $\preceq_K$ is determined in the following way:

$$E_{K,f} = \{(x, t) \in \mathbb{R}^{n+m} \mid f(x) \preceq_K t\}.$$

Show that

a) if the function f is K-convex, then its Lebesgue sets are convex;

b) the function f is K-convex, if and only if $E_{K,f}$ is a convex set.

26) The maximum of two (or more) K-convex functions is K-convex, but the situation is much more complicated than in the scalar case. Note that a, $b \in \mathbb{R}^m$ in the general case should not have a maximum relative to $\preceq_K$. In other words, there should not be $c \in \mathbb{R}^m$ such that $a \preceq_K c$, $b \preceq_K c$ and

$$a \preceq_K d,\ b \preceq_K d \Rightarrow c \preceq_K d.$$

Thus, the maximum of f_1 and f_2

$$f(x) = \max_K\{f_1(x), f_2(x)\}$$

is determined only in the case where for every x the functions $f_1(x)$ and $f_2(x)$ have a maximum. Show that when a maximum of two K-convex functions exists, then it is K-convex.

27) Let $g(x, y)$ be a K-convex function with respect to x and y, let C be a convex set and let for each x the set

$$\{g(x, y) \mid y \in C\}$$

have a minimal element (relative to $\preceq_K$), which we denote by $f(y)$. Show that the function f is K-convex.

5

Sub-gradient and Sub-differential of Finite Convex Function

By theorem 3.21, any convex function has derivatives with respect to all directions at the interior points of the domain of definition. At the same time, its partial derivatives, as well as gradient, may not exist. However, for convex functions the notions of sub-gradient and sub-differential (set of sub-gradients) can be introduced. These generalizations of the concepts of gradient and differential are used in the theory of non-smooth convex optimization problems.

In this chapter, we deal with finite-valued convex functions $f : \mathbb{R}^n \to \mathbb{R}$ exclusively.

5.1. Concepts of sub-gradient and sub-differential

Recall that for a differentiable at a point $\hat{x}$ convex function $f : \mathbb{R}^n \to \mathbb{R}^1$, the inequality

$$f(x) - f(\hat{x}) \geqslant \langle f'(\hat{x}), x - \hat{x} \rangle \quad \text{for all} \quad x \in X, \qquad [5.1]$$

holds true, which means that the graph of the function f lies not lower than the tangential hyperplane at the point $(\hat{x}, f(\hat{x}))$. According to the following definition, a sub-gradient is an arbitrary vector that can be substituted to inequality [5.1] instead of $f'(\hat{x})$.

DEFINITION 5.1.– *Let f be a function on the set $X \subset \mathbb{R}^n$. A vector $a \in \mathbb{R}^n$ is called the* sub-gradient *of the function f at point $\hat{x} \in X$, if*

$$f(x) - f(\hat{x}) \geqslant \langle a, x - \hat{x} \rangle \quad \textit{for all} \quad x \in X. \qquad [5.2]$$

The set of all sub-gradients is called the sub-differential *of the function f at point $\hat{x}$. It is denoted by $\partial f(\hat{x})$.*

DEFINITION 5.2.– *Let f be a function on the set $X \subset \mathbb{R}^n$. A vector $a \in \mathbb{R}^n$ is called the* super-gradient *of the function f at point $\hat{x} \in X$, if*

$$f(x) - f(\hat{x}) \leqslant \langle a, x - \hat{x} \rangle \quad \textit{for all} \quad x \in X.$$

The set of all super-gradients is called the super-differential *of the function f at point $\hat{x}$. It is denoted by $\overline{\partial} f(\hat{x})$.*

Relation [5.2] means that the graph

$$G_f = \{(x, \beta) \in X \times \mathbb{R} \mid f(x) = \beta)\}$$

of the function f lies not lower than the graph

$$H = \{(x, \beta) \in \mathbb{R}^n \times \mathbb{R} \mid l(x) = \beta\}$$

of the linear function $l(x) = f(\hat{x}) + \langle a, x - \hat{x} \rangle$. In this case, H is called the *support hyperplane* to the graph of the function f at the point $(\hat{x}, f(\hat{x}))$. Note that this is completely in line with the general concept of the support hyperplane, since in the notations introduced before we have $H = H_{p\alpha}$, where $p = (-a, 1)$, $\alpha = f(\hat{x}) - \langle a, \hat{x} \rangle$ and $(\hat{x}, f(\hat{x})) \in H_{p\alpha}$ while $G_f \subset H_{p\alpha}$. Moreover, H is a support hyperplane to the epigraph

$$\text{epi } f = \{(x, \beta) \in X \times \mathbb{R} \mid f(x) \leqslant \beta\}$$

of the function f at the point $\hat{x}$.

For a function of one real argument, the sub-gradient is the tangent of the angle of inclination of the support line (that is, the support hyperplane for $n = 1$), just as the derivative is the tangent of the tilt angle of the tangent line.

From geometric considerations, it follows that for a convex function f on a convex set $X \subset \mathbb{R}$, the following formula holds true:

$$\partial f(\hat{x}) = [f'_-(\hat{x}), f'_+(\hat{x})], \qquad [5.3]$$

where $f'_-(\hat{x})$ is the left-hand side derivative, and $f'_+(\hat{x})$ is the right-hand side derivative of the function f at the point $\hat{x}$. In other words, the support lines at point $(\hat{x}, f(\hat{x}))$ all take the intermediate values between the left-hand and right-hand tangents at this point.

For example, the sub-differential of the function $f(x) = |x|$ on $\mathbb{R}$ is of the form

$$\partial f(x) = \begin{cases} \{-1\}, & x < 0, \\ [-1, 1], & x = 0, \\ \{1\}, & x > 0. \end{cases}$$

The sub-differential of the function $f(x) = \|x\|$ on $\mathbb{R}^n$ is of the form

$$\partial f(x) = \begin{cases} \left\{\frac{x}{\|x\|}\right\}, & x \neq 0, \\ U_1(0), & x = 0, \end{cases}$$

where $U_1(0)$ is a ball of unit radius centered at zero. It follows from relation [5.3] that the convex function of a real argument is differentiable at a point $\hat{x}$ (that is $f'_-(\hat{x}) = f'_+(\hat{x}) = f'(\hat{x})$) if and only if its sub-differential at the point $\hat{x}$ contains only one element $f'(\hat{x})$.

5.2. Properties of sub-differential of convex function

First of all, we will show that the concepts of sub-gradient and sub-differential are closely related to the concept of a convex function.

THEOREM 5.1.– Let a function f be defined on a convex set $X \subset \mathbb{R}^n$. Then:

1) if the function f is convex on X, then its sub-gradient at any point $\hat{x} \in \operatorname{ri} X$ exists, that is $\partial f(\hat{x}) \neq \varnothing$, and $\partial f(\hat{x})$ is a closed convex set;

2) if $\partial f(x) \neq \varnothing$ for all $x \in X$, then the function f is convex on X.

PROOF.– 1) We apply theorem 3.22 on separating linear functions in the case where $X_1 = X$, $X_2 = \{\hat{x}\}$, $f_1(x) = f(x)$ for all $x \in X_1$, $f_2(\hat{x}) = f(\hat{x})$ (here $\operatorname{ri} X_2 = \{\hat{x}\}$, hence $\operatorname{ri} X_1 \cap \operatorname{ri} X_2 = \hat{x} \neq \varnothing$). So there is a linear function $l(x) = \langle a, x\rangle + b$ such that $f(x) \geqslant \langle a, x\rangle + b$ for all $x \in X$, and $\langle a, \hat{x}\rangle + b \geqslant f(\hat{x})$. Add these inequalities and get [5.2], that is $a \in \partial f(\hat{x})$. Closedness and convexity of $\partial f(\hat{x})$ follows directly from definition 5.2.

2) For all $x^1, x^2 \in X$, $\lambda \in [0,1]$ take $\bar{x} = \lambda x^1 + (1-\lambda)x^2 \in X$. By the condition of the theorem, there exists $a \in \partial f(\bar{x})$. Then

$$f(x^1) - f(\bar{x}) \geqslant \langle a, x^1 - \bar{x}\rangle$$

$$f(x^2) - f(\bar{x}) \geqslant \langle a, x^2 - \bar{x}\rangle$$

Multiply the first of the inequalities by λ, multiply the second inequality by $(1-\lambda)$ and add them. We come to inequality [3.1] from definition of the convex function (compare with the proof of theorem 3.6 for $\theta = 0$). □

REMARK 5.1.– At the relative boundary point, the subgradient of the convex function does not necessarily exist. An example can serve the function $f(x) = -\sqrt{1-x^2}$ on $X = [-1, 1]$. At points $\hat{x} = \pm 1$, the tangential to its graph (the only candidate for the support line) is in the vertical position and therefore cannot pass under the graph. In the following theorem, we establish two formulas that associate the sub-differential of a convex function with its directional derivatives (see theorem 3.21).

THEOREM 5.2.– Let f be a convex function on a convex set $X \subset \mathbb{R}^n$ and let $\hat{x} \in \operatorname{ri} X$. Then:

$$\partial f(\hat{x}) = \{a \in \mathbb{R}^n \mid f'(\hat{x}; h) \geqslant \langle a, h\rangle \quad \forall h \in \operatorname{Lin} X\}, \qquad [5.4]$$

$$f'(\hat{x}; h) = \max_{a \in \partial f(\hat{x})} \langle a, h\rangle \quad \forall h \in \operatorname{Lin} X. \qquad [5.5]$$

PROOF.– Denote the right side in [5.4] by Y. Let $a \in \partial f(\hat{x})$. Then for all $h \in \operatorname{Lin} X$, taking into account the definition and as well as the fact that the set A from [3.18] is non-empty, we have

$$f'(\hat{x}; h) = \lim_{\alpha \to 0+} \frac{f(\hat{x} + \alpha h) - f(\hat{x})}{\alpha} \geqslant \lim_{\alpha \to 0+} \frac{\langle a, \alpha h\rangle}{\alpha} = \langle a, h\rangle,$$

that is $a \in Y$. Let now $a \in Y$. Then for all $x \in X$, using formula [3.19] with $h = x - \hat{x} \in \operatorname{Lin} X$ and $\alpha = 1$, we get

$$f(x) - f(\hat{x}) \geqslant f'(\hat{x}; h) \geqslant \langle a, h\rangle = \langle a, x - \hat{x}\rangle$$

that is $a \in \partial f(\hat{x})$, which proves [5.4].

To prove the second relation [5.5], fix $h \in \operatorname{Lin} X$. If $h = 0$, then equality [5.5] is trivial: both its parts are zero. Let $h \neq 0$. Making use of formula [5.4], to prove [5.5] it is enough to show that

$$f'(\hat{x}; h) \leqslant \langle a, h\rangle \qquad [5.6]$$

for some $a \in \partial f(\hat{x})$. We again use theorem 3.22 on separating linear functions, where $X_1 = X$, $X_2 = \{x \in \mathbb{R}^n \mid x = \hat{x} + \alpha h,\ \alpha \geqslant 0\}$, $f_1(x) = f(x)$ for $x \in X_1$ and $f_2(x) = f(\hat{x}) + \frac{\|x - \hat{x}\|}{\|h\|} f'(\hat{x}; h)$ for $x \in X_2$. Since $\hat{x} \in \operatorname{ri} X$, then $\operatorname{ri} X_1 \cap \operatorname{ri} X_2 \neq \varnothing$. For all $x \in X_1 \cap X_2$, that is $x = \hat{x} + \alpha h \in X$ for some $\alpha \geqslant 0$, making use of formula [3.19], we get

$$f_1(x) = f(x) \geqslant f(\hat{x}) + \alpha f'(\hat{x}; h) = f_2(x).$$

In addition, $f_2(\lambda x^1 + (1-\lambda)x^2) = \lambda f_2(x^1) + (1-\lambda) f_2(x^2)$ for $x^1, x^2 \in X_2$, $\lambda \in [0, 1]$, that is the function f_2 is linear on X_2 (but not on $\mathbb{R}^n$). Hence, it is concave on X_2. Thus, conditions of theorem 3.22 are fulfilled and there exists a linear function $l(x) = \langle a, x\rangle + b$, such that the relations [3.21] and [3.22] hold true, that is

$$f(x) \geqslant \langle a, x\rangle + b \quad \forall x \in X,$$

$$\langle a, \hat{x} + \alpha h \rangle + b \geqslant f(\hat{x}) + \alpha f'(\hat{x}; h) \quad \forall \alpha \geqslant 0.$$

When $\alpha = 0$, we have $a \in \partial f(\hat{x})$ (see proof of theorem 5.1). Multiplying the last inequality by $1/\alpha$, $\alpha > 0$ and directing α to $+\infty$, we get [5.6]. □

Note that for the function f of one real argument, formula [5.4] is of the form [5.3], since $f'_+(\hat{x}) = f'(\hat{x}; 1)$ and $f'_-(\hat{x}) = -f'(\hat{x}; -1)$.

Formula [5.5] is a generalization to arbitrary convex functions of the formula

$$f'(\hat{x}; h) = \langle f'(\hat{x}), h \rangle \qquad [5.7]$$

which holds true for any function differentiable at point $\hat{x} \in \mathbb{R}^n$. This follows from the next theorem, which establishes the connection between the concept of sub-gradient and sub-differential with the notion of differentiability.

THEOREM 5.3.– Let f be a convex function on a convex set $X \subset \mathbb{R}^n$, and let $\hat{x} \in \operatorname{int} X$. Then:

1) if the function f is differentiable at point $\hat{x}$, then $\partial f(\hat{x}) = \{f'(\hat{x})\}$, that is the gradient $f'(\hat{x})$ is the only sub-gradient of the function f at point $\hat{x}$;

2) if $\partial f(\hat{x}) = \{a\}$, that is the sub-gradient of the function f at point $\hat{x}$ is the only one possible, then the function f is differentiable at the point $\hat{x}$, and $a = f'(\hat{x})$.

PROOF.– 1) Let $a \in \partial f(\hat{x})$. Note that $\operatorname{Lin} X = \mathbb{R}^n$ because $\hat{x} \in \operatorname{int} X$. It follows from [5.4] and [5.7] that for all $h \in \mathbb{R}^n$, we have $\langle f'(\hat{x}), h \rangle \geqslant \langle a, h \rangle$, that is $\langle f'(\hat{x}) - a, h \rangle \geqslant 0$. If we take $h = a - f'(\hat{x})$, then we will have $-\|f'(\hat{x}) - a\|^2 \geqslant 0$, that is $a = f'(\hat{x})$. Hence $\partial f(\hat{x}) = \{f'(\hat{x})\}$.

2) Denote by $U_r = U_r(0)$ a ball of radius r centered at zero. Since $\hat{x} \in \operatorname{int} X$, then $\hat{x} + U_r \subset \operatorname{int} X$ for some $r > 0$. Consider the function

$$\varphi(\alpha, h) = \frac{f(\hat{x} + \alpha h) - f(\hat{x})}{\alpha} - \langle a, h \rangle, \qquad [5.8]$$

where $\alpha \in (0, 1], h \in U_r$. Since $\partial f(\hat{x}) = \{a\}$, then formula [5.5] is of the form: $f'(\hat{x}; h) = \langle a, h \rangle$. By theorem 3.21 for arbitrary fixed $h \in U_r$, the function $\varphi(\alpha, h)$ is monotonically non-decreasing on $(0, 1]$ and converges to 0 as $\alpha \to 0+$. In this case, taking into account theorem 3.20, for an arbitrary fixed $\alpha \in (0, 1]$, the function $\varphi(\alpha, h)$ is continuous with respect to h on the compact set U_r. Therefore, according to the known Dini criterion, the indicated convergence is uniform on U_r, that is, $\forall \varepsilon > 0 \, \exists \delta \in (0, 1]$ such that $0 \leqslant \varphi(\alpha, h) \leqslant \varepsilon$ for all $\alpha \in (0, \delta), h \in U_r$. For arbitrary $h \in U_{\delta r}$, take $\alpha' = \frac{\|h\|}{r}, h' = \frac{rh}{\|h\|}$. Then $\alpha' \in (0, \delta)$, $h' \in U_r$ and $0 \leqslant \varphi(\alpha', h') \leqslant \varepsilon$. If we right these inequalities in more details and making use of [5.8], we get

$$0 \leqslant \frac{f(\hat{x} + h) - f(\hat{x}) - \langle a, h \rangle}{\|h\|} \leqslant \frac{\varepsilon}{r},$$

and

$$\lim_{h \to 0} \frac{f(\hat{x}+h) - f(\hat{x}) - \langle a, h \rangle}{\|h\|} = 0.$$

This means that f is differentiable at the point $\hat{x}$, and $a = f'(\hat{x})$ (take $h = \alpha e^j$ with $\alpha \to 0$, where by e^j we denote the jth unit vector in $\mathbb{R}^n$). □

REMARK 5.2.– The super-differentials of a concave function f have properties similar to those of sub-differentials of a convex function, since the function $g = -f$ is convex. For example:

– $\overline{\partial} f(\hat{x})$ is a non-empty closed convex set;

– $f'(\hat{x}; h) = \min\limits_{a \in \overline{\partial} f(\hat{x})} \langle a, h \rangle, \quad \forall h \in \operatorname{Lin} X.$

5.3. Sub-differential mapping

Denote by $\Pi(\mathbb{R}^m)$ the set of all non-empty subsets of the space $\mathbb{R}^m$. A mapping $F : X \to \Pi(\mathbb{R}^m)$, associating with each point $x \in X \subset \mathbb{R}^n$ a non-empty set $F(x) \subset \mathbb{R}^m$, is called multivalued or point-set mapping.

A multivalued mapping is also called a set-valued or many-valued mapping.

If for each $x \in X$, the set $F(x)$ consists of one element, then the mapping is called single valued.

If f is a convex function on an open convex set $X \subset \mathbb{R}^n$, then by theorem 5.1 at arbitrary point $x \in X$, the subdifferential $\partial f(x)$ (a subset of $\mathbb{R}^n$) is not empty. In this case, the multivalued mapping is given in a natural way: $\partial f : X \to \Pi(\mathbb{R}^n)$. This multivalued mapping is called *subdifferential mapping*.

Here are some concepts from the theory of multivalued mappings.

DEFINITION 5.3.– *Let $X \subset \mathbb{R}^n$. A multivalued mapping is called:*

1) closed, if from the condition

$$x^k \in X,\ y^k \in F(x^k),\ x^k \to x \in X,\ y^k \to y \in \mathbb{R}^m$$

it follows that $y \in F(x)$;

2) locally bounded, if from the conditions

$$x^k \in X,\ y^k \in F(x^k),\ x^k \to x \in X$$

it follows that the sequence y^k is bounded;

3) convex-valued, if $F(x)$ is a convex set for all $x \in X$;

4) monotone, if $m = n$ and

$$\langle y^1 - y^2, x^1 - x^2 \rangle \geqslant 0,$$

$$\forall x^1, x^2 \in X,\ \forall y^1 \in F(x^1),\ \forall y^2 \in F(x^2).$$

THEOREM 5.4.– Let f be a convex function on an open convex set $X \subset \mathbb{R}^n$. Then its subdifferential mapping $\partial f\colon X \to \Pi(\mathbb{R}^n)$ is closed, locally bounded, convex and monotone.

PROOF.–

1) *Closedness* of ∂f. Let

$$x^k \in X,\ a^k \in \partial f(x^k),\ x^k \to x \in X,\ a^k \to a.$$

By definition of $\partial f(x^k)$, we have

$$f(x') - f(x^k) \geqslant \langle a^k, x' - x^k \rangle \quad \forall x' \in X. \qquad [5.9]$$

with $f(x^k) \to f(x)$, since the function f is continuous on X (theorem 3.20). Passing to the limit in [5.9], we get

$$f(x') - f(x) \geqslant \langle a, x' - x \rangle \quad \forall x' \in X,$$

that is $a \in \partial f(x)$.

2) *Local boundedness* of ∂f. Let

$$x^k \in X,\ a^k \in \partial f(x^k),\ x^k \to x \in X.$$

Then the inequality [5.9] holds true. Suppose that the sequence $\{a^k\}$ is not bounded. We can assume that $\|a^k\| \to \infty$ and $\frac{a^k}{\|a^k\|} \to \bar{a} \neq 0$. Divide both sides of [5.9] by $\|a^k\|$ and, passing to the limit, we get

$$0 \geqslant \langle \bar{a}, x' - x \rangle \quad \forall x' \in X.$$

Since the set X is open, we can take $x' = x + \alpha \bar{a}$ for sufficiently small $\alpha > 0$. Hence $\bar{a} = 0$. This contradicts the previously mentioned $\bar{a} \neq 0$. Hence the sequence $\{a^k\}$ is bounded.

3) *Convexity* of ∂f. The convexity of the set $\partial f(x)$ for arbitrary $x \in X$ has already been noted in theorem 5.1.

4) *Monotonicity* of ∂f (compare with theorem 3.18). Let $x^1, x^2 \in X$, $a^1 \in \partial f(x^1)$, $a^2 \in \partial f(x^2)$. Then

$$f(x^2) - f(x^1) \geqslant \langle a^1, x^2 - x^1 \rangle, \quad f(x^1) - f(x^2) \geqslant \langle a^2, x^1 - x^2 \rangle.$$

Adding these inequalities, we get $\langle a^1 - a^2, x^1 - x^2 \rangle \geqslant 0$. □

It can be shown that a single-valued mapping is continuous if and only if it is closed and locally bounded as multivalued mapping with single-pointed sets of images. Hence, it follows from theorems 5.3 and 5.4 that the following theorem holds true.

THEOREM 5.5.– Let f be a differentiable convex function on an open convex set $X \subset \mathbb{R}^n$. Then the function f is continuously differentiable on X. That is, its gradient mapping f': $X \to \mathbb{R}^n$ is continuous.

This theorem shows how effective the notion of a subdifferential is even when studying differentiable convex functions.

For the mapping F_1: $X \to \Pi(\mathbb{R}^n)$, F_2: $X \to \Pi(\mathbb{R}^n)$, we will write

$$F_1 \cap F_2 \neq \varnothing,\ F_1 \subset F_2,\ F_1 = F_2,$$

when

$$F_1(x) \cap F_2(x) \neq \varnothing,\ F_1(x) \subset F_2(x),\ F_1(x) = F_2(x) \quad \text{for all} \quad x \in X.$$

Denote by $\operatorname{conv} F : X \to \Pi(\mathbb{R}^n)$ the convex hull of the mapping $F : X \to \Pi(\mathbb{R}^n)$, that is, a mapping that maps each point $x \in X$ to the set $\operatorname{conv} F(x)$. It is clear that $\operatorname{conv} F = F$, if F is a convex-valued mapping.

The following theorem will allow us to obtain a number of important properties of sub-differential mappings.

THEOREM 5.6.– Let X be an open set in $\mathbb{R}^n$, let F_1: $X \to \Pi(\mathbb{R}^n)$ be a closed locally bounded mapping, let F_2: $X \to \Pi(\mathbb{R}^n)$ be a monotone mapping and let $F_1 \cap F_2 \neq \varnothing$. Then $\operatorname{conv} F_2 \subset \operatorname{conv} F_1$.

PROOF.– Let $x \in X$ and $y \in F_2(x)$. Let us show that $y \in \operatorname{conv} F_1(x)$. Suppose it is not true, that is $y \notin \operatorname{conv} F_1(x)$. From the conditions on F_1, it follows that the set $F_1(x)$ is compact. Then $\operatorname{conv} F_1(x)$ is convex compact. It follows from the Minkowski theorem that there exists a vector $p \in \mathbb{R}^n$ such that

$$\langle y' - y, p \rangle < 0,\ \forall y' \in \operatorname{conv} F_1(x). \qquad [5.10]$$

Consider the sequence of points $x^k = x + p/k$ $(k = 1, 2, \ldots)$. Since X is an open set, then $x^k \in X$ for sufficiently large k. Under the condition for arbitrary k, there exists $y^k \in F_1(x^k) \cap F_2(x^k)$. From the locally boundedness of F_1, it follows that the sequence $\{y_k\}$ is bounded. Let $y^k \to y'$. Then $y' \in F_1(x)$, since F_1 is closed. Since F_2 is a monotone mapping, then $\langle y^k - y, x^k - x\rangle \geqslant 0$. This contradicts to condition [5.10]. That is why $y \in \operatorname{conv} F_1(x)$, that is $F_2 \subset \operatorname{conv} F_1$. Hence $\operatorname{conv} F_2 \subset \operatorname{conv} F_1$. □

Making use of theorems 5.4 and 5.6, we obtain the following result.

THEOREM 5.7.– Let f be a convex function on an open convex set $X \subset \mathbb{R}^n$ and let $F : X \to \Pi(\mathbb{R}^n)$ be a multivalued mapping. Then:

1) if the mapping F is closed, locally bounded and $F \cap \partial f \neq \varnothing$, then $\partial f \subset \operatorname{conv} F$;

2) if the mapping F monotone on $F \cap \partial f \neq \varnothing$, then $\operatorname{conv} F \subset \partial f$;

3) if the mapping F is closed, locally bounded and monotone on $F \cap \partial f \neq \varnothing$, then $\operatorname{conv} F = \partial f$;

4) if the mapping F is closed on $F \subset \partial f$, then $\operatorname{conv} F = \partial f$;

5) if the mapping F is monotone and $\partial f \subset F$, then $\partial f = F$.

Each of these statements in its own way characterizes properties of sub-differential mappings. For example, statement (5) means that these mappings are maximal in the class of monotone mappings. Statement (4) means that they are minimal in the class of closed and convex-valued mappings.

5.4. Calculus rules for sub-differentials

In Chapter 5, we describe the most common operations on convex functions that result in convex functions. In this section, we show how to calculate sub-differentials of these resulting convex functions. The importance of calculus rules increases in convex analysis because some operations on convex functions (max operation, for example) destroy differentiability but preserve convexity of functions.

THEOREM 5.8.– Let $f_1, \ldots, f_m$ be convex functions on an open convex set $X \subset \mathbb{R}^n$, $f_i \colon X \to \mathbb{R}^1$, and let $\alpha_1 \geq 0, \ldots, \alpha_m \geq 0$ be non-negative numbers. Then the sub-differential of the function $f(x) = \sum_{i=1}^{m} \alpha_i f_i(x)$ is of the form

$$\partial f(x) = \sum_{i=1}^{m} \alpha_i \partial f_i(x), \quad x \in X. \qquad [5.11]$$

Note that in the case when the functions $f_1, \ldots, f_m$ are differentiable, this formula has the form

$$f'(x) = \sum_{i=1}^{m} \alpha_i f_i'(x).$$

PROOF.– Denote the right part of the equality [5.11] by $F(x)$. As a result, there is a multivalued mapping $F : X \to \Pi(\mathbb{R}^n)$. From the closedness, the local boundednes and the convexity of the mappings ∂f_i (theorem 5.3), it follows that the mapping F is closed, convex-valued and $F \subset \partial f$.

Actually, let $x \in X$ and $a \in F(x)$. Then

$$a = \sum_{i=1}^{m} \alpha_i a^i, \; a^i \in \partial f_i, \; i = 1, \ldots, m.$$

For arbitrary $i = 1, \ldots, m$, by definition of $\partial f_i(x)$, we have

$$f_i(x') - f_i(x) \geqslant \langle a^i, x' - x \rangle \quad \forall x' \in X.$$

Hence, after multiplying by α_i and adding by i, we get

$$f(x') - f(x) \geqslant \langle a, x' - x \rangle \quad \forall x' \in X,$$

that is $a \in \partial f(x)$. As a result, we get $F \subset \partial f$. Then from statement (4) of theorem 5.7 it follows that $\partial f = \operatorname{conv} F = F$. In other words, [5.11] holds true. □

THEOREM 5.9.– Let X be an open convex set in $\mathbb{R}^n$, let Y be a compact and let $\varphi(x, y)$ be a function on $X \times Y$, $\varphi\colon X \times Y \to \mathbb{R}^1$, which is convex with respect to $x \in X$ for all $y \in Y$ and continuous with respect to all arguments on $X \times Y$. Then the sub-differential of the function $f(x) = \max_{y \in Y} \varphi(x, y)$, $x \in X$, is of the form

$$\partial f(x) = \operatorname{conv}\left(\bigcup_{y \in Y(x)} \partial\varphi(x, y) \right), \quad x \in X, \qquad [5.12]$$

where $Y(x) = \{y \in Y \mid \varphi(x, y) = f(x)\}$, and $\partial\varphi(x, y)$ is the sub-differential of the function $\varphi(x, y)$ with respect to X for a fixed $y \in Y$.

PROOF.– Denote by $F(x)$ the expression that is in [5.12] under the sign conv. Show that the mapping $F : X \to \Pi(\mathbb{R}^n)$ is closed. Let

$$x^k \in X, \; a^k \in F(x^k), \; x^k \to x \in X, \; a^k \to a.$$

Then $a^k \in \partial\varphi(x^k, y^k)$ for $y^k \in Y(x^k)$ such that $y^k \in Y$ and $\varphi(x^k, y^k) = f(x^k)$. Since Y is a compact, then we can assume that $y^k \to y \in Y$. Herewith $\varphi(x^k, y^k) \to$

$\varphi(x, y)$ and $f(x^k) \to f(x)$, since φ is continuous due to the conditions of the theorem, and f is continuous due to the conditions of theorem 3.20. Then $\varphi(x, y) = f(x)$, that is $y \in Y(x)$. By definition of $\partial\varphi(x^k, y^k)$, we have

$$\varphi(x', y^k) - \varphi(x^k, y^k) \geqslant \langle a^k, x' - x^k\rangle, \ \forall x' \in X.$$

Passing to the limit, we obtain

$$\varphi(x', y) - \varphi(x, y) \geqslant \langle a, x' - x\rangle, \ \forall x' \in X,$$

that is $a \in \partial\varphi(x, y), y \in Y(x)$. Therefore, $a \in F(x)$, which proves the closure of F.

Let now $x \in X$ and $a \in F(x)$, that is $a \in \partial\varphi(x, y)$ for some $y \in Y(x)$. Then $f(x) = \varphi(x, y)$ and for arbitrary $x' \in X$, we have

$$f(x') - f(x) \geqslant \varphi(x', y) - \varphi(x, y) \geqslant \langle a, x' - x\rangle,$$

that is $a \in \partial f(x)$. Therefore $F \subset \partial f$.

Making use statement (4) of theorem 5.7, we get $\partial f = \operatorname{conv} F$. This means that relation [5.12] holds true. □

REMARK 5.3.– It is clear from the proof of the theorem that the compactness condition on Y can be replaced by the weaker condition of boundedness of Y, if we additionally require that $\varphi(x^k, y^k) \to -\infty$, when $x^k \in X$, $y^k \in Y$, $x^k \to x \in X$, $y^k \to \overline{Y}\backslash Y$. This form of the theorem is more convenient to apply.

It is useful to reformulate the theorem for a maximum finite number of convex functions.

COROLLARY 5.1.– Let $f_1, \ldots, f_m$ be convex functions on an open convex set X, $f_i \colon X \to \mathbb{R}^1$. The sub-differential of the function $f(x) = \max\limits_{i=1,\ldots,m} f_i(x)$ is of the form

$$\partial f(x) = \operatorname{conv}\left\{\bigcup_{i \in I(x)} \partial f_i(x)\right\}, x \in X, \qquad [5.13]$$

where $I(x) = \{i \mid f_i(x) = f(x)\}$.

Making use of theorem 2.7 and the convexity of the set $\partial f_i(x)$, we can show that the formula [5.13] can be written in the form

$$\partial f(x) = \left\{a \in \mathbb{R}^n \,\middle|\, a = \sum_{i \in I(x)} \lambda_i a^i, a^i \in \partial f_i(x), \lambda_i \geqslant 0, \sum_{i \in I(x)} \lambda_i = 1\right\} \qquad [5.14]$$

In particular, in the case where the functions $f_1, \dots, f_m$ are differentiable on X, we have

$$\partial f(x) = \left\{ a \in \mathbb{R}^n \,\middle|\, a = \sum_{i \in I(x)} \lambda_i f_i'(x), \lambda_i \geqslant 0, \sum_{i \in I(x)} \lambda_i = 1 \right\}. \qquad [5.15]$$

To prove the following theorem, two lemmas are needed.

LEMMA 5.1.– Let $A_1, \dots, A_m$ be convex sets in $\mathbb{R}^n$, let P be a convex set in $\mathbb{R}^m$ and let $P \subset \mathbb{R}^m_+$. Then the set

$$A = \bigcup_{p \in P} \left(\sum_{i=1}^{m} p_i A_i \right),$$

where $p = (p_1, \dots, p_m)$ is convex.

PROOF.– Let $a, b \in A$, $\alpha \in (0, 1)$. We have

$$a = \sum_{i=1}^{m} p_i a^i, \quad b = \sum_{i=1}^{m} q_i b^i$$

for some $p, q \in P$, $a^i, b^i \in A_i$ $(i = 1, \dots, m)$. Take $r = \lambda p + (1 - \lambda) q \in P$ and $I = \{ i \,|\, r_i > 0 \}$. Note that $p_i = q_i = 0$ if $r_i = 0$. Then

$$\lambda a + (1 - \lambda) b = \sum_{i=1}^{m} (\lambda p_i a^i + (1 - \lambda) q_i b^i) =$$

$$\sum_{i \in I} r_i \left(\frac{\lambda p_i}{r_i} a^i + \frac{(1 - \lambda) q_i}{r_i} b_i \right) \in \sum_{i \in I} r_i A_i \subset A,$$

that is, A is convex. □

The following lemma is an analogue of the statement that the gradient of a monotonically non-decreasing differentiable function is non-negative.

LEMMA 5.2.– Let φ be a monotonically non-decreasing convex function on an open convex set $U \subset \mathbb{R}^m$. Then $\partial \varphi(u) \subset \mathbb{R}^m_+ \quad \forall u \in U$.

PROOF.– Let $p \in \partial \varphi(u)$, that is

$$\varphi(u') - \varphi(u) \geqslant \langle p, u' - u \rangle \quad \forall u' \in U$$

Taking $u' = u - \alpha e^i$, where e^i is the ith unit vector in $\mathbb{R}^m$, and $\alpha > 0$ is so small that $u' \in U$, we get $0 \geqslant \varphi(u') - \varphi(u) \geqslant -\alpha p_i$. Hence $p_i \geqslant 0, \forall i = 1, \dots, m$, that is $p \geqslant 0$. □

THEOREM 5.10.– Let $g_1, \dots, g_m$ be convex functions on an open convex set $X \subset \mathbb{R}^n$, let $g = (g_1, \dots, g_m)$ be a vector-valued function constructed from them, let φ be a monotonically non-decreasing convex function on an open convex set $U \subset \mathbb{R}^m$ and let $g(X) \subset U$. Then the sub-differential of the function $f(x) = \varphi(g(x))$ is of the form

$$\partial f(x) = \bigcup_{p \in \partial\varphi(u)} \left(\sum_{i=1}^{m} p_i \partial g_i(x) \right), \ x \in X, \qquad [5.16]$$

where $u = g(x)$.

PROOF.– Denote the right-hand side of [5.16] by $F(x)$. It follows from the closedness and local boundedness of the mappings ∂g_i and $\partial\varphi$ that the mapping F is closed. From the previous lemmas, as well as from the convexity of ∂g_i, $\partial\varphi$ follows the convexity of F. Let us verify that $F \subset \partial f$. Let $x \in X$ and $a \in F(x)$, that is $a = \sum_{i=1}^{m} p_i a^i$ for some $p \in \partial\varphi(u)$, where $u = g(x)$, and some $a^i \in \partial g_i(x)$, $i = 1, \dots, m$. Taking into account that $p \geqslant 0$ (lemma 5.2), for all $x' \in X$ we have

$$f(x') - f(x) = \varphi(g(x')) - \varphi(g(x)) \geqslant \langle p, g(x') - g(x) \rangle$$

$$= \sum_{i=1}^{m} p_i (g_i(x') - g_i(x)) \geqslant \sum_{i=1}^{m} p_i \langle a_i, x' - x \rangle = \langle a, x' - x \rangle,$$

that is, $a \in \partial f(x)$. Hence $F \subset \partial f$. Now we can use statement (4) of theorem 5.7. □

In particular, if the function φ is differentiable at the point $u = g(x)$, then the formula [5.16] is of the form

$$\partial f(x) = \sum_{i=1}^{m} \frac{\partial\varphi}{\partial u_i}(u) \partial g_i(x). \qquad [5.17]$$

If, moreover, the functions $g_1, \dots, g_m$are differentiable at the point x, then we get the well-known formula for the gradient of superposition of differentiable functions

$$f'(x) = \sum_{i=1}^{m} \frac{\partial\varphi}{\partial u_i}(u) g'_i(x).$$

The proof of the following theorem is similar.

THEOREM 5.11.– Let φ be a convex function on an open convex set $U \subset \mathbb{R}^m$, let A be a matrix of dimension $m \times n$, let $b \in \mathbb{R}^m$ and let the set $X = \{x \in \mathbb{R}^n \mid Ax + b \in U\}$ be non-empty. Then sub-differential of the function $f(x) = \varphi(Ax + b)$ is of the form

$$\partial f(x) = \partial\varphi(u) A \stackrel{def}{=} \{a \in \mathbb{R}^n \mid a = pA, \, p \in \partial\varphi(u)\},$$

where $u = Ax + b$.

EXAMPLE 5.1.– Find the sub-differential of the function

$$f(x) = \max\{x, x^2\}.$$

This function is determined by the formula

$$f(x) = \begin{cases} x, & x \in [0, 1]; \\ x^2, & x \in (-\infty, 0) \cup (1, \infty). \end{cases}$$

Hence by [5.14]

$$\partial f(x) = \begin{cases} 2x, & x < 0; \\ [0,1], & x = 0; \\ 1, & 0 < x < 1; \\ [1,2], & x = 1; \\ 2x, & x > 1. \end{cases}$$

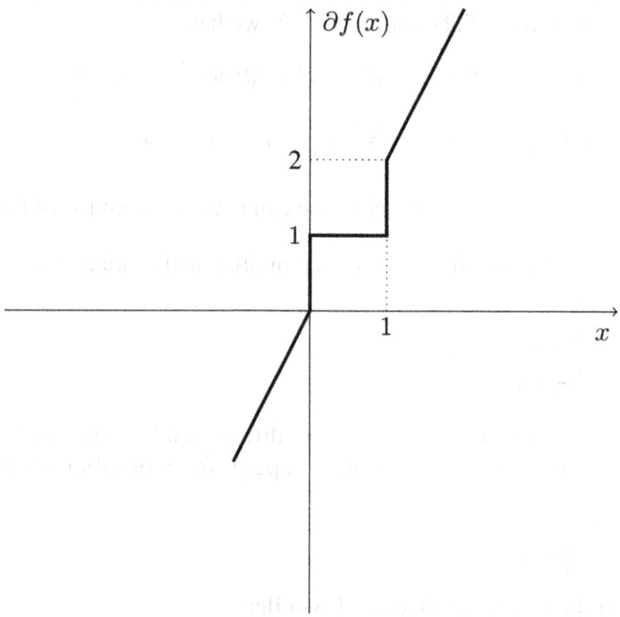

Figure 5.1. *Example 5.1*

EXAMPLE 5.2.– Let $f_0(x)$ be a convex function on an open convex set $X \subset \mathbb{R}^n$, and let $q \geqslant 1$ be a fixed number. Then the function

$$f(x) = [\max\{0, f_0(x)\}]^q$$

is convex. Let us calculate its sub-differential. Take $g(x) = \max\{0, f_0(x)\}$. From [5.17], it follows that

$$\partial f(x) = q[g(x)]^{q-1}\partial g(x),\ x \in X,$$

where by [5.14]

$$\partial g(x) = \begin{cases} \partial f_0(x), & f_0(x) > 0; \\ \{0\}, & f_0(x) < 0; \\ \{a \,|\, a = \lambda a',\ 0 \leqslant \lambda \leqslant 1,\ a' \in \partial f_0(x)\}, & f_0(x) = 0. \end{cases}$$

For $q = 1$, we have $\partial f(x) = \partial g(x)$. The answer is received.

Let $q > 1$. Note that $\partial f(x) = \{0\}$ for $f_0(x) \leqslant 0$ and, therefore, $g(x) = 0$. With this in mind, we can write

$$\partial f(x) = q[g(x)]^{q-1}\partial f_0(x) = q[\max\{0, f_0(x)\}]^{q-1}\partial f_0(x).$$

In particular, if the function f is differentiable at the point $x \in X$, then, by statement (1) of theorem 5.3, the sub-differential $\partial f_0(x)$ consists of one element $\forall x \in \mathbb{R}^n$. Then, according to statement (2) of theorem 5.3, the function f is differentiable at the point x, moreover

$$f'(x) = q[\max\{0, f_0(x)\}]^{q-1}\partial f_0(x).$$

EXAMPLE 5.3.– Find the sub-differential of the function

$$f(x) = \max_{1 \le i \le n} |x_i|,\ x = (x_1, \dots, x_n) \in \mathbb{R}^n.$$

This function can be represented in the form $f(x) = \max_{1 \le i \le n}\{f_i(x)\}$, where $f_i(x) = |x_i|$. We have

$$\partial f_i(x) = \begin{cases} -e_i, & x_i < 0; \\ +e_i, & x_i > 0; \\ \operatorname{conv}\{-e_i, +e_i\}, & x_i = 0, \end{cases}$$

where $e_i = (0, \dots, 0, 1, 0, \dots, 0)$ are unit vectors. Then by [5.14]

$$\partial f(x) = \operatorname{conv}\{w_i | i \in I(x)\},$$

$$I(x) = \{i | f(x) = |x_i|, i = 1, \dots, n\},$$

$$w_i = \begin{cases} -e_i, & x_i \le 0; \\ +e_i, & x_i \ge 0. \end{cases}$$

At point $x = 0$, we have

$$\partial f(0) = \operatorname{conv}\{\pm e_i | i = 1, \ldots, n\}.$$

For example, for the function

$$f(x_1, x_2) = \max\{|x_1|, |x_2|\}$$

we have

$$\partial f(x_1, x_2) = \begin{cases} (0,1), & x_1 > 0, x_2 > x_1; \\ \operatorname{conv}\{(0,1),(1,0)\}, & x_1 > 0, x_2 = x_1; \\ (1,0), & x_1 > 0, -x_1 < x_2 < x_1; \\ \operatorname{conv}\{(0,-1),(1,0)\}, & x_1 > 0, x_2 = -x_1; \\ (0,-1), & x_1 > 0, x_2 < -x_1; \\ (0,1), & x_1 < 0, x_2 > -x_1; \\ \operatorname{conv}\{(-1,0),(0,1)\}, & x_1 < 0, x_2 = -x_1; \\ (-1,0), & x_1 < 0, x_1 < x_2 < -x_1; \\ \operatorname{conv}\{(-1,0),(0,-1)\}, & x_1 > 0, x_2 = x_1; \\ (0,-1), & x_1 > 0, x_2 < x_1; \\ \operatorname{conv}\{(-1,0),(0,-1),(1,0),(0,1)\}, & x_1 = 0, x_2 = 0. \end{cases}$$

EXAMPLE 5.4.– Consider the piecewise affine function

$$f(x) = \max_{i=1,\ldots,m} \{\langle a_i, x\rangle - b_i\}, \; x \in \mathbb{R}^n,$$

where $a_i \in \mathbb{R}^n, b_i \in \mathbb{R}$. This function has the form

$$f(x) = \max_{i=1,\ldots,m} \{f_i(x)\}, \; x \in \mathbb{R}^n,$$

where $f_i(x) = \langle a_i, x\rangle - b_i, i = 1, \ldots, m$. Set

$$I(x) = \{i \colon f_i(x) = f(x)\}.$$

Making use of the representation

$$f(x + th) = f(x) + t \max\{\langle a_i, h\rangle, i \in I(x)$$

for small $t > 0$, we get

$$f'(x, h) = \max\{\langle a_i, h\rangle, i \in I(x)\}$$

and therefore the sub-differential

$$\partial f(x) = \operatorname{conv}\{a_i | i \in I(x)\}.$$

EXAMPLE 5.5.– To illustrate the calculus rules, consider the function $f\colon \mathbb{R}^p \times \mathbb{R}^q \to \mathbb{R}$ defined by

$$f(x_1, x_2) = f_1(x_1) + f_2(x_2),$$

where $f_1(x_1)$ and $f_2(x_2)$ are convex functions on $\mathbb{R}^p$ and $\mathbb{R}^q$ correspondingly. Define on $\mathbb{R}^p \times \mathbb{R}^q$ the following functions:

$$\tilde{f}_1(x_1, x_2) = f_1(x_1), \quad \tilde{f}_2(x_1, x_2) = f_2(x_2).$$

Their sub-differentials are as follows:

$$\partial\tilde{f}_1(x_1, x_2) = \partial f_1(x_1) \times \{0\}, \quad \partial\tilde{f}_2(x_1, x_2) = \{0\} \times \partial f_2(x_2).$$

Therefore

$$\partial f_1(x_1, x_2) = \partial f_1(x_1) \times \{0\} + \{0\} \times \partial f_2(x_2) = \partial f_1(x_1) \times \partial f_2(x_2).$$

From these considerations, it follows that the sub-differential of the function

$$f(x) = \sum_{i=1}^{n} |x_i|, \; x = (x_1, \dots, x_n) \in \mathbb{R}^n.$$

is of the form

$$\partial f(x_1, \dots, x_n) = \partial|x_1| \times \cdots \times \partial|x_n|$$

and

$$\partial f(0, \dots, 0) = \{(u_1, \dots, u_n) \in \mathbb{R}^n \colon \max_{1 \le i \le n} |u_i| \le 1\}.$$

For example, for the function

$$f(x_1, x_2) = |x_1| + |x_2|$$

we have

$$\partial f(0,0) = \{(u_1, u_2)\colon |u_1| \le 1, |u_2| \le 1\}.$$

At other points

$$\partial f(x_1, x_2) = \begin{cases} (1,1), & x_1 > 0, x_2 > 0; \\ \operatorname{conv}\{(1,-1),(1,1)\}, & x_1 > 0, x_2 = 0; \\ (1,-1), & x_1 > 0, x_2 < 0; \\ (-1,1), & x_1 < 0, x_2 > 0; \\ \operatorname{conv}\{(-1,-1),(-1,1)\}, & x_1 < 0, x_2 = 0; \\ (-1,-1), & x_1 < 0, x_2 < 0; \\ \operatorname{conv}\{(-1,1),(1,1)\}, & x_1 = 0, x_2 > 0; \\ \operatorname{conv}\{(-1,-1),(1,-1),(1,1),(-1,1)\}, & x_1 = 0, x_2 = 0; \\ \operatorname{conv}\{(-1,-1),(1,-1)\}, & x_1 = 0, x_2 < 0. \end{cases}$$

EXAMPLE 5.6.– Let $f(x) = \|x\|$ be an arbitrary norm in $\mathbb{R}^n$. It is a positive (except at 0) closed sublinear function and its sublevel set

$$B = \{x \in \mathbb{R}^n \colon \|x\| \leq 1\}$$

is the unit ball associated with the norm. It is symmetric, convex, compact set containing the origin as an interior point. The norm $f(x) = \|x\|$ is gauge of B. And it is a support function of the set

$$B^* = \{s \in \mathbb{R}^n \colon \langle s, x\rangle \leq \|x\| \text{ for all } x \in \mathbb{R}^n\}$$

which is also symmetric, convex, compact containing the origin as an interior point. The support function of the unit ball B and the gauge of B^* are the same function $\|x\|^*$ defined by

$$\|x\|^* = \max\{\langle s, x\rangle \colon \|x\| \leq 1\}.$$

The support function of the unit ball B^* and the gauge of B are the same function $\|x\|$ defined by

$$\|x\| = \max\{\langle s, x\rangle \colon \|s\|^* \leq 1\}.$$

Note that the symmetric relation (Cauchy–Bunyakovsky inequality)

$$\langle s, x\rangle \leq \|s\|^*\|x\| \quad \text{for all } s \in \mathbb{R}^n,\ x \in \mathbb{R}^n$$

expresses the duality correspondence between the following two Banach spaces:

$$(\mathbb{R}^n, \|\cdot\|) \text{ and } (\mathbb{R}^n, \|\cdot\|^*).$$

The sub-differential of the norm $f(x) = \|x\|$ at zero coincides with a closed unit ball of the conjugate space. Actually

$$f(x) - f(0) = \|x\| - \|0\| = \|x\| \geq \langle s, x - 0\rangle = \langle s, x\rangle.$$

It follows from the definition that

$$\partial\|0\| = \{s \in \mathbb{R}^n \colon \|s\|^* \leq 1\} = B^*.$$

If $x \neq 0$, then the sub-differential

$$\partial\|x\| = \{s \in \mathbb{R}^n \colon \ \|s\|^* = 1,\ \langle s, x\rangle = \|x\|\}.$$

Actually, if $\langle s, x\rangle = \|x\|, \|s\|^* = 1$, then $\|z\| \geqslant \langle s, z\rangle$ for all $z \in \mathbb{R}^n$. Hence

$$f(z) - f(x) = \|z\| - \|x\| \geqslant \langle s, z - x\rangle,$$

that is $s \in \partial\|x\|$.

On the contrary, if $s \in \partial f(x) = \partial\|x\|$, then

$$-\|x\| = \|0\| - \|x\| \geqslant \langle s, 0 - x\rangle = -\langle s, x\rangle,$$

so $\|x\| \leq \langle s, x\rangle$. On the other hand

$$\|x\| = f(2x) - f(x) = \|2x\| - \|x\| \geqslant \langle s, 2x - x\rangle = \langle s, x\rangle,$$

hence $\|x\| = \langle s, x\rangle$.

Next, for all $z \in \mathbb{R}^n$, $\lambda > 0$

$$\|\lambda z + x\| - \|x\| \geqslant \langle s, \lambda z\rangle,$$

$$\left\|z + \frac{x}{\lambda}\right\| - \frac{1}{\lambda}\|x\| \geqslant \langle s, z\rangle,$$

where for $\lambda \to \infty$ it follows that

$$\|z\| \geqslant \langle s, z\rangle$$

for all $z \in \mathbb{R}^n$, that is $\|s\|^* \leqslant 1$. But since $\langle s, x\rangle = \|x\|$, we have $\|s\|^* = 1$. Hence the sub-differential of the norm $f(x) = \|x\|$ is of the form

$$\partial\|x\| = \begin{cases} \{s \in \mathbb{R}^n : \|s\|^* \leqslant 1\}, & x = 0, \\ \{s \in \mathbb{R}^n : \|s\|^* = 1, \langle s, x\rangle = \|x\|\}, & x \neq 0. \end{cases}$$

EXAMPLE 5.7.– The sub-differential of the indicator function $\delta(x|A)$ is a non-empty set for each $x \in A$ (if $x \in A$, then $0 \in \partial\delta(x|A)$). By definition

$$\partial\delta(x|A) = \{a \in \mathbb{R}^n : \langle a, z - x\rangle \leqslant 0, \, z \in A\}.$$

We can see that $\partial\delta(x|A)$ is a cone. It is called the *cone of support functionals*, or the *normal cone* of the set A at point x and is denoted by $N(x|A)$. In particular, if $A = L$ is a subspace, then

$$\partial\delta(x|L) = N(x|L) = L^{\perp},$$

where $L^{\perp}$ is the annihilator of L.

5.5. Systems of convex and linear inequalities

THEOREM 5.12.– *(Fan's theorem)*. Let X be a convex set in $\mathbb{R}^n$, let $f_1(x), \dots, f_k(x)$ be convex functions on X and let $f_{k+1}(x), \dots, f_m(x)$ be linear functions on R^n. Assume that the system

$$f_i(x) < 0, \qquad i = 1, \dots, k; \qquad [5.18]$$

$$f_i(x) = 0, \quad i = k+1, \ldots, m, \tag{5.19}$$

has no solutions on X. Then there are numbers $y_1 \geq 0, \ldots, y_k \geq 0$, $y_{k+1}, \ldots, y_m$, which are not all equal to zero and such that

$$\sum_{i=1}^{m} y_i f_i(x) \geq 0 \quad \text{for all} \quad x \in X. \tag{5.20}$$

PROOF.– Consider the sets

$$U_1 = \{u \in \mathbb{R}^m \colon \exists x \in X \colon f_i(x) \leq u_i, i = 1, \ldots, k; f_i(x) = u_i, i = k+1, \ldots, m\},$$

$$U_2 = \{u \in \mathbb{R}^m \colon u_i < 0,\ i = 1, \ldots, k;\ u_i = 0, i = k+1, \ldots, m\}.$$

The fact that the systems [5.18] and [5.19] have no solutions on X means that $U_1 \cap U_2 = \emptyset$. It is easy to verify that the set U_1 is convex (here conditions of convexity and linearity are essential). The convexity of U_2 is obvious. Then, by theorem 2.21 on the separation of sets, there exists a non-zero vector $y \in \mathbb{R}^m$ such that

$$\langle y, u\rangle \geq \langle y, v\rangle \quad \text{for all} \quad u \in U_1, v \in U_2.$$

In other words

$$\sum_{i=1}^{m} y_i u_i \geq \sum_{i=1}^{k} y_i v_i \quad \text{for all} \quad u \in U_1,\ v_1 \leq 0, \ldots, v_k \leq 0. \tag{5.21}$$

Hence, for $v_1 \to -\infty, \ldots, v_k \to -\infty$, we obtain that $y_1 \geq 0, \ldots, y_k \geq 0$. For any $x \in X$, consider a vector $u \in \mathbb{R}^m$ with coordinates $u_i = f_i(x), i = 1, \ldots, m$. This vector $u \in U_1$. Substituting this vector in [5.21] with $v_1 = \cdots = v_k = 0$, we come to [5.20]. □

Under the conditions of theorem 5.12 the numbers $y_1, \ldots, y_k$, which correspond to inequalities [5.18], are nonnegative. The following theorem establishes the conditions under which the relation [5.20] takes place, where the number y_0, which corresponds to a strict inequality, is positive. Theorems of this type are called theorems with regularity conditions.

THEOREM 5.13.– Let X be a convex set in $\mathbb{R}^n$ and let $f_1(x), \ldots, f_k(x)$ be convex functions on X. Assume that the system

$$f_0(x) < 0, \tag{5.22}$$

$$f_i(x) < 0, \quad i = 1, \ldots, m, \tag{5.23}$$

has no solutions on X, but its subsystem [5.23] has solutions on X. Then there are numbers $y_0 > 0, y_1 \geq 0, \ldots, y_m \geq 0$, not all equal to zero, and such that

$$y_0 f_0(x) + \sum_{i=1}^{m} y_i f_i(x) \geq 0 \quad \text{for all} \quad x \in X. \tag{5.24}$$

PROOF.– By theorem 5.12, there are numbers $y_0 \geq 0, y_1 \geq 0, \ldots, y_m \geq 0$, which are not all equal to zero and such that [5.24] holds true. Suppose that $y_0 = 0$. In this case, among the numbers $y_1, \ldots, y_m$ there are positive numbers. Then for the point $x \in X$, which is a solution of the system [5.23], we have $\sum_{i=1}^{m} y_i f_i(x) < 0$. This contradicts to [5.24], where $y_0 = 0$. Hence $y_0 > 0$. □

Note that in the ratio [5.24], we can assume that $y_0 = 1$. This can be achieved by dividing all its terms by $y_0 > 0$.

The previously proved theorem 2.22 (Farkas theorem) is a theorem with regularity conditions for a system of homogeneous linear inequalities. Let' us rewrite it in the form and notations of theorem 5.12.

THEOREM 5.14.– Assume that the system

$$f_0(x) = \langle a_0, x \rangle < 0,$$

$$f_i(x) = \langle a_i, x \rangle \leq 0, \quad i = 1, \ldots, m,$$

has no solutions on $\mathbb{R}^n$. Then there are numbers $y_1 \geq 0, \ldots, y_m \geq 0$ such that

$$a_0 + \sum_{i=1}^{m} y_i a_i = 0,$$

that is

$$f_0(x) + \sum_{i=1}^{m} y_i f_i(x) = 0 \quad \text{for all} \quad x \in \mathbb{R}^n.$$

On the basis of this result, we can obtain a theorem with regularity conditions for a system of non-homogeneous linear inequalities.

THEOREM 5.15.– Assume that the system

$$f_0(x) = \langle a_0, x \rangle + b_0 < 0, \tag{5.25}$$

$$f_i(x) = \langle a_i, x \rangle + b_i \leq 0, \quad i = 1, \ldots, m, \tag{5.26}$$

has no solutions on $\mathbb{R}^n$, but its subsystem [5.26] has solutions on X. Then there are numbers $y_1 \geq 0, \ldots, y_m \geq 0$ such that

$$a_0 + \sum_{i=1}^{m} y_i a_i = 0, \quad b_0 + \sum_{i=1}^{m} y_i b_i \geq 0, \tag{5.27}$$

that is

$$f_0(x) + \sum_{i=1}^{m} y_i f_i(x) \geq 0 \quad \text{for all} \quad x \in \mathbb{R}^n.$$

PROOF.– Consider a system of homogeneous linear inequalities.

$$\langle a_0, h\rangle + b_0\lambda < 0, \qquad [5.28]$$

$$\langle a_i, h\rangle + b_i\lambda \le 0, \quad i = 1, \ldots, m, \qquad [5.29]$$

$$\langle 0, h\rangle - \lambda \le 0. \qquad [5.30]$$

Assume that the system has a solution $(h, \lambda) \in \mathbb{R}^n \times R$. By virtue of [5.30], we have $\lambda > 0$ or $\lambda = 0$. If $\lambda > 0$, then from [5.28] and [5.29] it follows that $x = h/\lambda$ is a solution of systems [5.25] and [5.26]. And this contradicts the condition of the theorem. If $\lambda = 0$, then $\langle a_0, h\rangle < 0$ and $\langle a_i, h\rangle \le 0$, $i = 1, \ldots, m$. Fix a solution x of the system [5.26]. Then

$$\langle a_0, x + \alpha h\rangle + b_0 = (\langle a_0, x\rangle + b_0) + \alpha\langle a_0, h\rangle < 0$$

at a sufficiently large $\alpha > 0$ and

$$\langle a_i, x + \alpha h\rangle + b_i \le 0, \quad i = 1, \ldots, m$$

for arbitrary $\alpha \ge 0$.

Consequently, systems [5.25] and [5.26] have a solution of the form $x + \alpha h$, which again contradicts the condition of the theorem. Consequently, systems [5.28]–[5.30] have no solution. Then, by theorem 5.14, there exist numbers $y_1 \ge 0, \ldots, y_m \ge 0, y_{m+1} \ge 0$ such that

$$(a_0, b_0) + \sum_{i=1}^{m} y_i(a_i, b_i) + y_{m+1}(0, -1) = 0,$$

that is, [5.27] holds true. □

Now we formulate more general results for systems of linear inequalities and equations.

THEOREM 5.16.– Let X be a polyhedron in $\mathbb{R}^n$, and let $f_0(x), f_1(x), \ldots, f_k(x)$, $f_{k+1}(x), \ldots, f_m(x)$ be linear functions on $\mathbb{R}^n$. Assume that the system

$$f_0(x) < 0, \qquad [5.31]$$

$$f_i(x) \le 0, \quad i = 1, \ldots, k; \qquad [5.32]$$

$$f_i(x) = 0, \quad i = k + 1, \ldots, m, \qquad [5.33]$$

has no solutions on X, but its subsystems [5.32] and [5.33] have solutions on X. Then there are numbers $y_1 \ge 0, \ldots, y_k \ge 0, y_{k+1}, \ldots, y_m$ such that

$$f_0(x) + \sum_{i=1}^{m} y_i f_i(x) \ge 0 \quad \text{for all} \quad x \in X. \qquad [5.34]$$

PROOF.– The polyhedron X can be represented on the form

$$X = \{x \in \mathbb{R}^n | g_j(x) \leq 0, \quad j = 1, \ldots, s\},$$

where $g_1(x), \ldots, g_s(x)$ are linear functions. Consider the system

$$\begin{aligned} &f_0(x) < 0, \\ &f_i(x) \leq 0, \ \ i = 1, \ldots, k; \\ &f_i(x) \leq 0, \ \ i = k+1, \ldots, m; \\ -&f_i(x) \leq 0, \ i = k+1, \ldots, m; \\ &g_j(x) \leq 0, \ \ j = 1, \ldots, s. \end{aligned}$$

This system satisfies conditions of theorem 5.15. Therefore, there exist non-negative numbers $y_1 \geq 0, \ldots, y_k \geq 0, u_{k+1} \geq 0, \ldots, u_m \geq 0, v_{k+1} \geq 0, \ldots, v_m \geq 0, z_1 \geq 0, \ldots, z_s \geq 0$ such that

$$f_0(x) + \sum_{i=1}^{k} y_i f_i(x) + \sum_{i=k+1}^{m} (u_i - v_i) f_i(x) + \sum_{j=1}^{s} z_j g_j(x) \geq 0$$

for all $x \in \mathbb{R}^n$. Take $y_i = u_i - v_i$ for $i = k+1, \ldots, M$. Then

$$f_0(x) + \sum_{i=1}^{m} y_i f_i(x) \geq -\sum_{j=1}^{s} z_j g_j(x) \geq 0$$

for $x \in X$. □

Based on the obtained results, we prove two theorems. The first one is a direct generalization of theorem 5.16, which is associated with the substitution in [5.31] and [5.32] of some linear functions by convex functions. In this case, the corresponding inequalities in [5.32] become strict.

THEOREM 5.17.– Let X be a polyhedron in $\mathbb{R}^n$, let $f_0(x), f_1(x), \ldots, f_l(x)$ be convex functions that are defined on a relatively open set $U \supset X$ and let $f_{l+1}(x), \ldots, f_k(x), f_{k+1}(x), \ldots, f_m(x)$ be linear functions on $\mathbb{R}^n$. Suppose that the system

$$f_0(x) < 0, \qquad [5.35]$$

$$f_i(x) < 0, \quad i = 1, \ldots, l; \qquad [5.36]$$

$$f_i(x) \leq 0, \quad i = l+1, \ldots, k; \qquad [5.37]$$

$$f_i(x) = 0, \quad i = k+1, \ldots, m, \qquad [5.38]$$

has no solutions on X, but its subsystems [5.36]–[5.38] have solutions on X. Then there are numbers $y_1 \geq 0, \ldots, y_k \geq 0, y_{k+1}, \ldots, y_m$ such that

$$f_0(x) + \sum_{i=1}^{m} y_i f_i(x) \geq 0 \quad \text{for all} \quad x \in X. \qquad [5.39]$$

PROOF.– Consider the set

$$V = \{x \in X | f_i(x) \leq 0, \; i = l+1, \ldots, k; \; f_i(x) = 0, \; i = k+1, \ldots, m\}. \quad [5.40]$$

According to conditions of the theorem, systems [5.35] and [5.36] have no solutions on V, but subsystem [5.36] has solutions on X. Then by theorem 5.13 there exist numbers $y_1 \geq 0, \ldots, y_l \geq 0$ such that the function

$$f(x) = f_0(x) + \sum_{i=1}^{l} y_i f_i(x) \geq 0 \quad \text{for all} \quad x \in V. \quad [5.41]$$

Moreover, the function f is convex on U. In addition, $V \subset U = \operatorname{ri} U$. Hence $\operatorname{ri} U \cap \operatorname{ri} V = \operatorname{ri} V$. Then by theorem 3.22 (for $X_1 = U, X_2 = V, f_1 = f, f_2 = 0$) there exists a linear function $g(x)$ such that

$$f(x) \geq g(x) \quad \text{for all} \quad x \in U. \quad [5.42]$$

$$g(x) \geq 0 \quad \text{for all} \quad x \in V. \quad [5.43]$$

It follows from [5.40] and [5.43] that the system formed from the inequality $g(x) < 0$ and relations [5.37] and [5.38] have no solutions on X. Then by theorem 5.16, there exist numbers $y_{l+1} \geq 0, \ldots, y_k \geq 0, y_{k+1}, \ldots, y_m$ such that

$$g(x) = \sum_{i=l+1}^{m} y_i f_i(x) \geq 0 \quad \text{for all} \quad x \in X. \quad [5.44]$$

Since $X \subset U$, then from inequalities [5.41], [5.42] and [5.44] the inequality [5.39] follows. □

LEMMA 5.3.– Let X_1 and X_2 be convex sets in $\mathbb{R}^n$, and let $\operatorname{ri} X_1 \cap X_2 \neq \emptyset$ $X_2 \subset \operatorname{aff} X_1$. Then $\operatorname{ri} X_1 \cap \operatorname{ri} X_2 \neq \emptyset$.

PROOF.– Consider points $x \in \operatorname{ri} X_1 \cap X_2$, $y \in \operatorname{ri} X_2$, $z^\alpha = x + \alpha(y - x), \alpha \in R$. Then $z^\alpha \in \operatorname{ri} X_2$ for all $\alpha \in (0, 1]$ (theorem 2.14). At the same time, since $x \in \operatorname{ri} X_1$, $y \in \operatorname{ri} X_2 \subset \operatorname{aff} X_1$, we have $z^\alpha \in \operatorname{aff} X_1$ for all α. But $\lim_{\alpha \to 0} z^\alpha = x$. Therefore $z^\beta \in X_1$ for sufficiently small β. Hence, $z^\alpha = (1 - \alpha/\beta)x + \alpha/\beta z^\beta \in \operatorname{ri} X_1$ for all $\alpha \in [0, \beta)$ (theorem 2.14). So $z^\alpha \in \operatorname{ri} X_1 \cap \operatorname{ri} X_2$ if $0 < \alpha < \min\{1, \beta\}$. □

The following theorem is a modification of the previous one. It is proved by means of theorems 5.13, 3.22 and 5.16.

THEOREM 5.18.– Let X be a convex set in $\mathbb{R}^n$, let $f_0(x), f_1(x), \ldots, f_l(x)$ be convex functions on X and let $f_{l+1}(x), \ldots, f_k(x), f_{k+1}(x), \ldots, f_m(x)$ be linear functions on

$\mathbb{R}^n$. Assume that the systems [5.35]–[5.38] have no solutions on X, but subsystems [5.36]–[5.38] have solutions on ri X. Then the assertion of theorem 5.17 holds true.

PROOF.– Together with the set V defined by condition [5.40], let us consider the set

$$V' = \{x \in \operatorname{aff} X \colon f_i(x) \leq 0, i = l+1, \ldots, k; f_i(x) = 0, i = k+1, \ldots, m\}. \quad [5.45]$$

Then $V = X \cap V'$. By theorem 5.13, there exist numbers $y_1 \geq 0, \ldots, y_l \geq 0$ such that the function f [5.41] is non-negative on V. It is convex on X. Moreover, $\operatorname{ri} X \cap V' \neq \emptyset$, $V' \subset \operatorname{aff} X$. That is why $\operatorname{ri} X \cap \operatorname{ri} V' \neq \emptyset$. Then by theorem 3.22 (for $X_1 = X, X_2 = V', f_1 = f, f_2 = 0$), there exists a linear function $g(x)$ such that

$$f(x) \geq g(x) \quad \text{for all} \quad x \in X. \quad [5.46]$$

$$g(x) \geq 0 \quad \text{for all} \quad x \in V'. \quad [5.47]$$

It follows from [5.41] and [5.46] that the system consisting of the inequality $g(x) < 0$ and relations [5.37] and [5.38] has no solutions on aff X. The affine set aff X is a polyhedron (theorem 2.5). Then by theorem 5.16 there exist numbers $y_{l+1} \geq 0, \ldots, y_k \geq 0, y_{k+1}, \ldots, y_m$ such that

$$g(x) = \sum_{i=l+1}^{m} y_i f_i(x) \geq 0 \quad \text{for all} \quad x \in \operatorname{aff} X.$$

From this and [5.41] and [5.46], we obtain [5.39]. □

REMARK 5.4.– The theorem cannot be limited to the condition that the systems [5.36]–[5.38] have a solution on X. An example is the system

$$\begin{aligned} f_0(x) &= x_1 - 1 < 0, \\ f_1(x) &= x_1 + x_2 - 2 \leq 0 \end{aligned}$$

on the set $X = \{x \in \mathbb{R}^2 | \, x_1 x_2 \geq 1, x_1 \geq 0, x_2 \geq 0\}$

5.6. Exercises

1) Let f be a convex function on a convex set X. Prove that $\partial f(\hat{x}) \cap \operatorname{Lin} X \neq \varnothing$ for arbitrary $\hat{x} \in \operatorname{ri} X$, i.e. the sub-gradient can always be selected from Lin X.

2) Let f be a convex function on a convex set X. Let X_0 be a compact subset of ri X. Show that f satisfies the Lipschitz condition on X_0, that is there exists a number L such that

$$|f(x^1) - f(x^2)| \leqslant L\|x^1 - x^2\| \quad \forall x^1, x^2 \in X_0.$$

Hint: show that the set $\bigcup_{x \in X_0} (\partial f(x) \cap \operatorname{Lin} X)$ (non-empty by virtue of Problem 1) is bounded and then use definition 5.2.

3) Prove that in theorems 5.8–5.11 the directional derivative of the function f is of the form

$$f'(x;h) = \sum_{i=1}^{m} \alpha_i f_i'(x;h), \quad f'(x;h) = \max_{y \in Y(x)} \varphi'(x,h,y),$$

where $\varphi'(x,h,y)$ is the derivative of the function $\varphi(x,y)$ as a function with respect to x in the direction of h for a fixed y;

$$f'(x;h) = \max_{p \in \partial\varphi(u)} \sum_{i=1}^{m} p_i g_i'(x;h), \quad u = g(x); \quad f'(x;h) = \varphi'(Ax+b, Ah).$$

4) Find subdifferentials and directional derivatives of the functions:

a) $f(x) = 2|x-1| + |x+2|,\ x \in \mathbb{R}$;

b) $f(x) = |x+3| + 3|x|,\ x \in \mathbb{R}$;

c) $f(x) = \max\{e^x, 1-x\},\ x \in \mathbb{R}$;

d) $f(x_1,x_2) = 3|x_1+1| + 2|x_2-3|,\ x_1,x_2 \in \mathbb{R}$;

e) $f(x_1,x_2) = 2(x_1-1)^2 + 3|x_2+2|,\ x_1,x_2 \in \mathbb{R}$;

f) $f(x_1,x_2) = \max\{ax_2 + (x_1-b)^2, 0\},\ x_1,x_2 \in \mathbb{R}$;

g) $f(x_1,x_2) = 3|x_1+2| - x_2 + x_2^2,\ x_1,x_2 \in \mathbb{R}$;

h) $f(x_1,x_2) = \max\{2|x_1| - x_2 + 1, |x_1+1| + 2|x_2-2|\},\ x_1,x_2 \in \mathbb{R}$;

i) $f(x_1,x_2) = x_1^2 + x_1x_2 + x_2^2 + |x_1 - x_2 - 2|,\ x_1,x_2 \in \mathbb{R}$;

j) $f(x_1,x_2) = x_1^2 + x_2^2 + 2\max\{|x_1|, |x_2|\},\ x_1,x_2 \in \mathbb{R}$;

k) $f(x_1,x_2) = x_1^2 + x_2^2 + \sqrt{(x_1-1)^2 + (x_2+2)^2},\ x_1,x_2 \in \mathbb{R}$;

l) $f(x_1,x_2,x_3) = |x_1 - x_2 + 3x_3|,\ x_1,x_2,x_3 \in \mathbb{R}$;

m) $f(x) = \max\{0, \langle a,x\rangle\},\ a,x \in \mathbb{R}^n$;

n) $f(x) = \max_{i=1,\dots,n} \{x_i\},\ x \in \mathbb{R}^n$;

o) $f(x) = \sum_{i=1}^{m} |\langle a_i,x\rangle - b_i|,\ x \in \mathbb{R}^n$;

p) $f(x) = e^{\|x\|},\ x \in \mathbb{R}^n$;

q) $f(x) = \|Ax - b\|,\ x \in \mathbb{R}^n$.

6

Constrained Optimization Problems

6.1. Differential conditions of optimality

Consider the minimization problem in a general form:

$$f(x) \to \min, \quad x \in X \subset \mathbb{R}^n. \qquad [6.1]$$

Let us introduce some definitions.

DEFINITION 6.1.– *Let $X \subset \mathbb{R}^n$. A vector $h \in \mathbb{R}^n$ determines a possible direction with respect to the set X at a point $\hat{x} \in X$ if $\hat{x} + \alpha h \in X$ for all sufficiently small $\alpha > 0$. We denote the set of all such vectors h by $V(\hat{x}, X)$.*

DEFINITION 6.2.– *A vector h determines a direction of decrease of a function f at a point $\hat{x} \in \mathbb{R}^n$ if $f(\hat{x} + \alpha h) < f(\hat{x})$ for all sufficiently small $\alpha > 0$. We denote the set of all such vectors h by $U(\hat{x}, f)$.*

LEMMA 6.1.– Let the function f be differentiable at a point $\hat{x} \in \mathbb{R}^n$. Then:

– if a vector h satisfies condition

$$\langle f'(\hat{x}), h \rangle < 0, \qquad [6.2]$$

then $h \in U(\hat{x}, f)$;

– if $h \in U(\hat{x}, f)$, then $\langle f'(\hat{x}), h \rangle \leq 0$.

PROOF.– Let condition [6.2] be satisfied. Then

$$f(\hat{x} + \alpha h) - f(\hat{x}) = \langle f'(\hat{x}), \alpha h \rangle + o(\alpha) = \alpha \left(\langle f'(\hat{x}), h \rangle + \frac{o(\alpha)}{\alpha} \right) < 0$$

for all sufficiently small $\alpha > 0$, that is $h \in U(\hat{x}, f)$.

Let $\langle f'(\hat{x}), h\rangle > 0$. Then h is a direction of increase of the function f at a point $\hat{x}$ and $h \notin U(\hat{x}, f)$. Consequently, from the condition $h \in U(\hat{x}, f)$, the inequality $\langle f'(\hat{x}), h\rangle \leq 0$ follows. □

Below is the condition of local optimality for problem [6.1], which does not require any assumptions about the set X and functions f.

THEOREM 6.1.– If $\hat{x}$ is a local solution of the problem [6.1], then

$$U(\hat{x}, f) \cap V(\hat{x}, X) = \emptyset.$$

PROOF.– Assume that $U(\hat{x}, f) \cap V(\hat{x}, X) \neq \emptyset$, that is, there exists a vector $h \in \mathbb{R}^n$ for which $f(\hat{x} + \alpha h) < f(\hat{x})$ and $\hat{x} + \alpha h \in X$ for all sufficiently small $\alpha > 0$. Then, in any small neighborhood of the point $\hat{x}$, there is a point $x = \hat{x} + \alpha h \in X$ such that $f(x) < f(\hat{x})$. This contradicts the definition of the local solution of the problem [6.1]. □

THEOREM 6.2.– Suppose that in the problem [6.1], the set X is convex and the function f is differentiable at a point $\hat{x} \in X$. Then:

1) if $\hat{x}$ is a local solution of the problem [6.1], then

$$\langle f'(\hat{x}), x - \hat{x}\rangle \geq 0 \quad \text{for all} \quad x \in X; \qquad [6.3]$$

2) if the function f is convex on X and condition [6.3] is satisfied, then $\hat{x}$ is a (global) solution of the problem [6.1].

PROOF.– 1) Assume that $\hat{x}$ is a local solution of problem [6.1] and [6.3] is not true, that is $\langle f'(\hat{x}), x - \hat{x}\rangle < 0$ for some $x \in X$. Take $h = x - \hat{x}$. Then $h \in U(\hat{x}, f)$ by lemma 6.1. At the same time, from the convexity of X for any $\alpha \in [0, 1]$, we have $\hat{x} + \alpha h = \alpha x + (1 - \alpha)\hat{x} \in X$, that is $h \in V(\hat{x}, X)$. Hence $U(\hat{x}, f) \cap V(\hat{x}, X) \neq \emptyset$. This contradicts the statement of theorem 6.1.

The validity of (2) follows from theorems 3.16 and 3.23. □

Thus, relation [6.3] is a necessary condition for the local extremum in the problem of minimization of a differentiable function on a convex set. For the convex optimization problem, this relation is also a sufficient condition for a global minimum.

Geometrically, condition [6.3] means that the gradient $f'(\hat{x})$ (if it is non-zero) forms a non-obtuse angle with a vector coming from the point $\hat{x}$ in any direction $x \in X$.

Condition [6.3] can be written in a more precise form in some special cases.

LEMMA 6.2.– Let $\hat{x} \in \operatorname{int} X$. In this case, condition [6.3] is equivalent to the following condition:

$$f'(\hat{x}) = 0, \quad \text{that is} \quad \frac{\partial f(\hat{x})}{\partial x_j} = 0, \quad j = 1, \ldots, n. \qquad [6.4]$$

PROOF.– For sufficiently small $\alpha > 0$, the point $x = \hat{x} \pm \alpha f'(\hat{x})$ belongs to X. Substituting x into [6.3], we obtain [6.4]. On the contrary, from [6.4] condition [6.3] follows. □

Thus, for the unconstrained optimization problem ($X = \mathbb{R}^n$), theorem 6.2 does not give any new results compared to known traditional results.

LEMMA 6.3.– Let the set X be of the form

$$X = \{x \in \mathbb{R}^n \mid a_j \le x_j \le b_j, j = 1, \ldots, n\}, \qquad [6.5]$$

where $-\infty \le a_j < b_j \le +\infty, j = 1, \ldots, n$ (if $a_j = -\infty$ or $b_j = +\infty$, then the corresponding sign of inequality in [6.5] should be considered as strict). Then condition [6.3] is equivalent to the condition

$$\frac{\partial f(\hat{x})}{\partial x_j} \begin{cases} = 0, \text{ if } a_j < \hat{x}_j < b_j; \\ \ge 0, \text{ if } \hat{x}_j = a_j \ne -\infty; \\ \le 0, \text{ if } \hat{x}_j = b_j \ne +\infty. \end{cases} \qquad [6.6]$$

for all $j = 1, \ldots, n$.

PROOF.– In the case where the set X is of the indicated form, condition [6.3] is equivalent to the following condition: for any $j = 1, \ldots, n$:

$$\frac{\partial f(\hat{x})}{\partial x_j}(x_j - \hat{x}_j) \ge 0 \quad \text{for all} \quad x_j \in [a_j, b_j].$$

And this is equivalent to [6.6]. □

Consider a special case of this assertion.

LEMMA 6.4.– Let the set X be of the form

$$X = \{x \in \mathbb{R}^n \mid x_j \ge 0, j = 1, \ldots, s\},$$

where $0 \le s \le n$ ($s = 0$ corresponds to $X = \mathbb{R}^n$). Then condition [6.3] is equivalent to the following conditions:

$$\frac{\partial f(\hat{x})}{\partial x_j} \ge 0; \quad \hat{x}_j \cdot \frac{\partial f(\hat{x})}{\partial x_j} = 0, \quad j = 1, \ldots, s; \qquad [6.7]$$

$$\frac{\partial f(\hat{x})}{\partial x_j} = 0, \quad j = s+1, \ldots, n. \qquad [6.8]$$

PROOF.– Note that X is a set of the form [6.5] with $a_j = 0$, $j = 1, \ldots, s$; $a_j = -\infty$, $j = s+1, \ldots, n$; $b_j = +\infty$, $j = 1, \ldots, n$. In this case, conditions [6.6] are equivalent to conditions [6.7] and [6.8]. □

LEMMA 6.5.– Let X be an affine set and let $L = X - \hat{x}$ be parallel to it subspace. Then condition [6.3] is equivalent to the following condition:

$$\langle f'(\hat{x}), h \rangle = 0 \quad \forall h \in L,$$

that is $f'(\hat{x})$ lies in the orthogonal complement to L.

In simple cases, the obtained results allow us to solve the problem [6.1] explicitly.

EXAMPLE 6.1.– Find all (local and global) solutions to the problem:

$$f(x_1, x_2) = 2x_1^2 + x_1x_2 + x_2^2 \to \min, \qquad -1 \leq x_1 \leq 1,\ x_2 \geq 2. \qquad [6.9]$$

According to the assertion of lemma 6.3, the following conditions have to be fulfilled:

$$\frac{\partial f(x_1, x_2)}{\partial x_1} = 4x_1 + x_2 \begin{cases} = 0, \text{ if } -1 < x_1 < 1; \\ \geq 0, \text{ if } x_1 = -1; \\ \leq 0, \text{ if } x_1 = 1. \end{cases} \qquad [6.10]$$

$$\frac{\partial f(x_1, x_2)}{\partial x_2} = x_1 + 2x_2 \begin{cases} = 0, \text{ if } x_2 > 2; \\ \geq 0, \text{ if } x_2 = 2. \end{cases} \qquad [6.11]$$

Now, in general, it is necessary to make six systems of pairs of relations combining relations [6.10]and [6.11]. For example, the first two systems have the form:

$$4x_1 + x_2 = 0,\ -1 < x_1 < 1,\ x_1 + 2x_2 = 0,\ x_2 > 2; \qquad [6.12]$$

$$4x_1 + x_2 = 0,\ -1 < x_1 < 1,\ x_1 + 2x_2 \geq 0,\ x_2 = 2. \qquad [6.13]$$

After that, you need to find solutions to each such system and test them for optimality. However, before that it is useful to conduct a qualitative analysis of the problem.

The function f is quadratic. Using the Sylvester criterion, we can make sure that the matrix of the second derivatives of the function f is positive definite. Consequently, the function f is strongly convex on $\mathbb{R}^2$. Therefore, the local and global solutions of the problem [6.9] coincide. The problem has a unique solution and only this solution can satisfy conditions [6.10] and [6.11]. So, not yet solving these six systems, we already know that only one of them have a solution, and it is a solution to the problem [6.9].

Now consider the system [6.12]. It is incompatible. Let us go to system [6.13]. Its solution is $x_1 = -1/2, x_2 = 2$. This is a solution to the problem [6.9]. Considering the other four systems does not make sense. As mentioned above, they cannot have solutions.

6.2. Sub-differential conditions of optimality

By theorem 6.2, relation [6.3] is a necessary and sufficient condition for optimality in a convex minimization problem with a differentiable function. In the following theorem, we formulate more general results, which also cover the case of non-differentiable functions.

THEOREM 6.3.– Let, in the optimization problem [6.1], the set X be convex and the function f be convex on a relatively open set U, which contains X. Then a point $\hat{x} \in X$ is a solution of the problem [6.1] if and only if there exists a vector $a \in \partial f(\hat{x})$ such that

$$\langle a, x - \hat{x} \rangle \geq 0 \quad \forall x \in X. \tag{6.14}$$

Here $\partial f(\hat{x})$ is the sub-differential of the function f, which is considered on U. In other words, the notation $a \in \partial f(\hat{x})$ means that

$$f(x) - f(\hat{x}) \geq \langle a, x - \hat{x} \rangle \quad \forall x \in U. \tag{6.15}$$

If the function f is differentiable at the point $\hat{x}$, then $\partial f(\hat{x}) = \{f'(\hat{x})\}$ and relation [6.4] transforms to relation [6.3].

PROOF.– Let $\hat{x}$ be a solution to the problem [6.1]. Since $X \subset U = \operatorname{ri} U$ and $\operatorname{ri} U \cap \operatorname{ri} X = \operatorname{ri} X \neq \emptyset$, then we apply theorem 3.22 on the separating linear function in the case where $X_1 = U$, $X_2 = X$, $f_1(x) = f(x) - f(\hat{x})$, $f_2(x) = 0$. According to the assertion of theorem 3.22, there exists a linear function $l(x) = \langle a, x \rangle + b$ such that

$$f(x) - f(\hat{x}) \geq \langle a, x \rangle + b \quad \forall x \in U, \tag{6.16}$$

$$\langle a, x \rangle + b \geq 0 \quad \forall x \in X. \tag{6.17}$$

Hence, for $\hat{x} = x$, we obtain $b = -\langle a, \hat{x} \rangle$. As a result, relation [6.16] transforms to relation [6.15], and relation [6.17] transforms to relation [6.14].

Sufficiency of conditions is obvious: if for some $a \in \partial f(\hat{x})$ [6.14] takes place, then from [6.15] it follows that $f(x) - f(\hat{x}) \geq 0$ for all $x \in X$, that is $\hat{x}$ is a solution to the problem [6.1]. □

REMARK 6.1.– Under the conditions of theorem 6.3, one would assume that the function f is defined only on X and, accordingly, in relation [6.15] instead of U take X. But in this case, the theorem becomes trivial: $a = 0$ always fits. Making use the above-mentioned theorem, we can apply the theory of sub-differentials in the study of convex optimization problems.

Note that lemmas 6.1–6.4 can be reformulated taking into account relation [6.14]. In particular, if $\hat{x} \in \operatorname{int} X$, then condition [6.14] is equivalent to the condition $a = 0$. In other words, in the assumptions of theorem 6.3 a point $\hat{x} \in \operatorname{int} X$ is a solution of the problem [6.1] only in the case where $0 \in \partial f(\hat{x})$. According to the earlier remark, this fact itself is trivial; however, it can be useful, if we can calculate the sub-differential $\partial f(\hat{x})$.

EXAMPLE 6.2.– Find all solutions to the problem:

$$f(x) = \|x\| - \langle c, x\rangle \to \min, \; x \in \mathbb{R}^n,$$

where c is a vector from $\mathbb{R}^n$.

The function f is convex on $\mathbb{R}^n$. Its sub-differential at the point $x = 0$ has the form $\partial f(0) = B_1(0) - c$. In all other points, the function f is differentiable and $f'(x) = x/\|x\| - c$. Inclusion $0 \in \partial f(0)$ means that $c \in B_1(0)$, that is $\|c\| \leq 1$. Equation $f'(x) = 0$ has a solution only when $\|c\| \leq 1$. In this case, if $\|c\| = 1$, then its solution is any point $\hat{x} = \lambda \cdot c$, where $\lambda > 0$.

So the answer is as follows: if $\|c\| < 1$, then $\hat{x} = 0$ is the unique solution of this problem; if $\|c\| = 1$, then any point $\hat{x} = \lambda \cdot c$, where $\lambda > 0$, will be a solution of the problem; if $\|c\| > 1$, then there are no solutions.

EXAMPLE 6.3.– Find solutions to the problem:

$$f(x_1, x_2) = x_1^2 + x_1x_2 + x_2^2 + 3|x_1 + x_2 - 2| \to \min .$$

The function $f(x) = f(x_1, x_2)$ is convex as a sum of two convex functions. Indeed, the function $g(x_1, x_2) = x_1^2 + x_1x_2 + x_2^2$ is convex, since the matrix of the second derivatives

$$g''(x) = \left(\frac{\partial g(x)}{\partial x_i \partial x_j}\right)^2_{i,j=1} = \begin{pmatrix} 2 & 1 \\ 1 & 2 \end{pmatrix}$$

is positive definite and does not depend on x. The function $h(x_1, x_2) = |x_1 + x_2 - 2|$ is also convex as a maximum of two linear functions. The necessary and sufficient condition of extremum of a convex unconstrained problem is

$$0 \in \partial f(\hat{x}) = \partial g(\hat{x}) + 3\partial h(\hat{x}).$$

Since the function $g(x)$ is differentiable, its sub-differential coincides with the derivative

$$\partial g(x) = (2x_1 + x_2, x_1 + 2x_2).$$

The sub-differential of the function $h(x_1, x_2) = |x_1 + x_2 - 2|$ is calculated by the formula

$$\partial h(x_1, x_2) = \begin{cases} (1,1), & x_1 + x_2 - 2 > 0; \\ (\alpha, \alpha), |\alpha| \leq 1, & x_1 + x_2 - 2 = 0; \\ (-1,-1), & x_1 + x_2 - 2 < 0. \end{cases}$$

Hence,

$$\partial f(x_1, x_2) = \begin{cases} (2x_1 + x_2 + 3, x_1 + 2x_2 + 3), & x_1 + x_2 - 2 > 0; \\ (2x_1 + x_2 + 3\alpha, x_1 + 2x_2 + 3\alpha), & x_1 + x_2 - 2 = 0; \\ (2x_1 + x_2 - 3, x_1 + 2x_2 - 3), & x_1 + x_2 - 2 < 0. \end{cases}$$

Consequently, the condition of the extremum $0 \in \partial f(x)$ is of the form

$$\begin{cases} 2x_1 + x_2 + 3 = 0, \\ x_1 + 2x_2 + 3 = 0, \\ x_1 + x_2 - 2 > 0; \end{cases}$$

$$\begin{cases} 2x_1 + x_2 + 3\alpha = 0, \\ x_1 + 2x_2 + 3\alpha = 0, \\ x_1 + x_2 - 2 = 0; \end{cases}$$

$$\begin{cases} 2x_1 + x_2 - 3 = 0, \\ x_1 + 2x_2 - 3 = 0, \\ x_1 + x_2 - 2 < 0. \end{cases}$$

In the first and third cases, there are no critical points, because the systems of conditions are incompatible. In the second case, we obtain a solution $x_1 = 1, x_2 = 1, \alpha = -1$.

So the answer is the following: $S_{\min} = 3$, $(x_1, x_2) = (1, 1)$.

6.3. Exercises

Find solutions to the following problems:

1) $4x_1^2 - x_1x_2 + 2x_2^2 \to \min, \quad 4 \leq x_1 \leq 8, \quad -1 \leq x_2 \leq 2$.

2) $ax_1^2 + bx_1x_2 + cx_2^2 \to \min, \quad -1 \leq x_1 \leq 1, \quad x_2 \geq 1$, where a, b, c are such numbers that $a > 0, 4ac > b^2$.

3) $ax_1^2 + x_1x_2 + x_2^2 \to \min, \quad 2 \leq x_1 \leq 3, \quad 3 \leq x_2 \leq 4$, where $a \in \mathbb{R}$.

4) $\frac{1}{2}\|x\|^2 - \langle c, x\rangle \to \min, \quad x \geq 0.$

5) $\|x\| - \langle c, x\rangle \to \min, \quad x \geq 0.$

6) $\frac{1}{2}\|x\|^2 + \|x - c\| \to \min, \quad x \in \mathbb{R}^n.$

7) $x_1^2 - x_1x_2 + x_2^2 + |x_1 - x_2 - 2| \to \min.$

8) $x_1^2 + x_2^2 + 4\max\{x_1, x_2\} \to \min.$

9) $x_1^2 + x_2^2 + 2\sqrt{(x_1 - a_1)^2 + (x_2 - a_2)^2} \to \min.$

10) $x_1^2 + x_2^2 + a|x_1 + x_2 - 1| \to \min.$

11) Specify the value of the parameter $a \in \mathbb{R}$ for which the point $(0, 0)$ is a solution of the problem

$$e^{a^2x_1} + e^{a^2x_2} + 2ax_1 - x_2 \to \min, \quad 0 \leq x_1 \leq 1, \quad -1 \leq x_2 \leq 0.$$

12) In problem [6.1], let the set X be convex and the function f have a derivative in any direction $h \in V(\hat{x}, X)$ at a point $\hat{x} \in X$, that is the value

$$f'(\hat{x}, h) = \lim_{\alpha \to +0} \frac{f(\hat{x} + \alpha h) - f(\hat{x})}{\alpha}$$

exists and it is finite. Show that $\langle f'(\hat{x}), h\rangle \geq 0$ for all $h \in V(\hat{x}, X)$ if $\hat{x}$ is a local solution of the problem [6.1].

6.4. Constrained optimization problems

6.4.1. *Principle of indeterminate Lagrange multipliers*

Consider the following constrained optimization problem:

$$\begin{array}{l} f(x) \to \min, \\ g_i(x) \leq 0, \quad i = 1, \ldots, k; \\ g_i(x) = 0, \quad i = k+1, \ldots, m; \\ x \in P \subset \mathbb{R}^n. \end{array} \qquad [6.18]$$

This problem can be reduced to the problem [6.1] if we define the admissible set X as

$$X = \{x \in P \,| g_i(x) \leq 0, i = 1, \ldots, k;\ g_i(x) = 0,\ i = k+1, \ldots, m\}. \qquad [6.19]$$

Also define the set

$$Q = \{y = (y_1, \dots, y_m) \in \mathbb{R}^m \mid y_i \geq 0, i = 1, \dots, k\}, \qquad [6.20]$$

consisting of all m-dimensional vectors in which first k coordinates are non-negative. In particular, $Q = \mathbb{R}^m$ if there are no inequality constraints, and $Q = R_+^m$ if there are no equality constraints ($k = m$). Define the *Lagrange function* of the problem [6.18]

$$L(x, y_0, y) = y_0 f(x) + \sum_{i=1}^{m} y_i g_i(x),$$

where $x \in P, y_0 \geq 0, y = (y_1, \dots, y_m) \in Q$. This function has the same form as in the case of the classical constrained optimization problem with differentiable functions. We will continue to use the notation

$$L'_x(x, y_0, y) = y_0 f'(x) + \sum_{i=1}^{m} y_i g'_i(x), \qquad [6.21]$$

for a vector of partial derivatives of the Lagrange function with respect to coordinates of the vector x, that is, with quantities

$$\frac{\partial L}{\partial x_j}(x, y_0, y) = y_0 \frac{\partial f}{\partial x_j}(x) + \sum_{i=1}^{m} y_i \frac{\partial g_i}{\partial x_j}(x), \quad j = 1, \dots, n.$$

THEOREM 6.4.– (*Principle of indeterminate Lagrange multipliers*) Suppose that in the problem [6.18] the set P is convex, the functions $f, g_1, \dots, g_k$ are differentiable at a point $\hat{x} \in X$ and the functions $g_{k+1}, \dots, g_m$ are differentiable in a neighborhood of the point $\hat{x}$. If $\hat{x}$ is a local solution of the problem [6.18], then there exist a number $\hat{y}_0 \geq 0$ and a vector $\hat{y} = (\hat{y}_1, \dots, \hat{y}_m) \in Q$, not all equal to zero and such that

$$\langle L'_x(\hat{x}, \hat{y}_0, \hat{y}), x - \hat{x}\rangle \geq 0 \quad \forall x \in P, \qquad [6.22]$$

$$\hat{y}_i g_i(\hat{x}) = 0, \quad i = 1, \dots, k. \qquad [6.23]$$

REMARK 6.2.– We will make a number of remarks on this theorem and the method of indeterminate Lagrange multipliers.

1) Any point $\hat{x} \in X$ satisfying conditions [6.22] and [6.23] for some $\hat{y}_0 \geq 0$, $\hat{y} \in Q$, $(\hat{y}_0, \hat{y}) \neq 0$, is called the *stationary point* of the problem [6.18]. The Lagrange principle states that under the assumptions of theorem 6.4 any local solution of the problem [6.18] is a stationary point. The sufficiency of conditions [6.22] and [6.23] is guaranteed under some additional assumptions only (see theorems 6.5 and 6.10).

2) The numbers $\hat{y}_0, \hat{y}_1, \dots, \hat{y}_m$ are called *Lagrange multipliers*. According to the definition of the set Q, the multipliers $\hat{y}_1, \dots, \hat{y}_k$, corresponding to the inequality

constraints, are non-negative, and the multipliers $\hat{y}_{k+1}, \ldots, \hat{y}_m$, which correspond to equality constraints, can be negative and positive. The Lagrange multipliers are defined up to a positive constant, that is, if the pair $(\hat{y}_0, \hat{y})$ satisfies conditions [6.22] and [6.23], then for any $\lambda > 0$ the pair $(\lambda\hat{y}_0, \lambda\hat{y})$ also satisfies conditions [6.22] and [6.23]. This allows us to consider only two cases in theorem 6.4: $\hat{y}_0 = 0$ or $\hat{y}_0 = 1$. Additional conditions that indicate the case $\hat{y}_0 = 1$ are called *regularity conditions*. The problem itself is called *regular* in this case. For such a problem, it is sufficient to consider the Lagrange function of the form

$$L(x, y) = L(x, 1, y) = f(x) + \sum_{i=1}^{m} y_i g_i(x), \qquad [6.24]$$

which is also called the *regular* Lagrange function. For a regular convex optimization problem, relations [6.22] and [6.23] are not only necessary, but also sufficient conditions of optimality.

THEOREM 6.5.– Suppose that in the problem [6.18] the set P is convex, the functions $f, g_1, \ldots, g_k$ are convex on P and differentiable at a point $\hat{x} \in X$ and the functions $g_{k+1}, \ldots, g_m$ are linear. If for $\hat{y}_0 = 1$ and some $\hat{y} \in Q$ conditions [6.22] and [6.23] hold true, then $\hat{x}$ is a global solution of the problem [6.18].

PROOF.– Since under conditions of the theorem, the function $L(x, \hat{y})$ is convex with respect to x on P; then from condition [6.22], it follows that $L(x, \hat{y})$ attains the minimum at the point $\hat{x}$, that is

$$L(\hat{x}, \hat{y}) \leq L(x, \hat{y}) \quad \forall x \in P.$$

Taking into account this fact, as well as [6.23], we have for any $x \in X$

$$f(\hat{x}) = f(\hat{x}) + \sum_{i=1}^{m} \hat{y}_i g_i(\hat{x}) = L(\hat{x}, \hat{y}) \leq L(x, \hat{y}) = f(x) + \sum_{i=1}^{m} \hat{y}_i g_i(x) \leq f(x).$$

□

Based on lemmas 6.2–6.4, we describe condition [6.22] more precisely in some special cases.

LEMMA 6.6.– Let the conditions of theorem 6.4 be satisfied. Then:

1) if $\hat{x} \in \operatorname{int} P$, then condition [6.22] is equivalent to the condition

$$L'_x(\hat{x}, \hat{y}_0, \hat{y}) = 0, \quad \text{that is} \quad \frac{\partial L}{\partial x_j}(\hat{x}, \hat{y}_0, \hat{y}) = 0, \; j = 1, \ldots, n; \qquad [6.25]$$

2) if the set P is of the form

$$P = \{x \in \mathbb{R}^n \,|a_j \leq x_j \leq b_j, \; j = 1, \ldots, n\},$$

where $-\infty \leq a_j < b_j \leq +\infty,\ j = 1, ..., n$, then condition [6.22] is equivalent to the condition: $\forall j = 1, \ldots, n$

$$\frac{\partial L}{\partial x_j}(\hat{x}, \hat{y}_0, \hat{y}) \begin{cases} = 0, & \text{if} \quad a_j < \hat{x}_j < b_j, \\ \geq 0, & \text{if} \quad \hat{x}_j = a_j \neq -\infty, \\ \leq 0, & \text{if} \quad \hat{x}_j = b_j \neq +\infty; \end{cases}$$

3) if the set P is of the form

$$P = \{x \in \mathbb{R}^n \,|x_j \geq 0,\, j = 1, \ldots, s\}, \qquad [6.26]$$

where $0 \leq s \leq n$, then condition [6.22] is equivalent to the condition:

$$\frac{\partial L}{\partial x_j}(\hat{x}, \hat{y}_0, \hat{y}) \geq 0, \quad \hat{x}_j \frac{\partial L}{\partial x_j}(\hat{x}, \hat{y}_0, \hat{y}) = 0, \quad j = 1, \ldots, s,$$

$$\frac{\partial L}{\partial x_j}(\hat{x}, \hat{y}_0, \hat{y}) = 0, \qquad j = s+1, \ldots, n.$$

For any point $\hat{x} \in X$, we define the sets

$$I(\hat{x}) = \{\, i \,|g_i(\hat{x}) = 0, \ \ 1 \leq i \leq k\},$$

$$S(\hat{x}) = I(\hat{x}) \cup \{k+1, \ldots, m\} = \{\, i \mid g_i(\hat{x}) = 0, \ \ 1 \leq i \leq m\}.$$

The inequality constraints with $i \in I(\hat{x})$ are called *active* at the point $\hat{x}$, while other conditions are called *passive*. Condition [6.23], which is sometimes called the *complementary slackness condition*, means that the Lagrange multipliers that correspond to passive inequality constraints must be equal to zero, that is $\hat{y}_i = 0$ for all $i \in \{1, \ldots, k\} \backslash I(\hat{x})$. From conditions [6.22] and [6.23], taking into account [6.21], we get

$$\left\langle \hat{y}_0 f'(\hat{x}) + \sum_{i \in S(\hat{x})} \hat{y}_i g_i'(\hat{x}), x - \hat{x} \right\rangle \geq 0 \quad \forall x \in P. \qquad [6.27]$$

On the contrary, from [6.27] we can always go to [6.22] and [6.23] if we consider $\hat{y}_i = 0$ for $i \in \{1, \ldots, k\} \backslash I(\hat{x})$.

Let us explain the geometric content of the Lagrangian principle, when there is no equality constraints and restriction $x \in P$ ($k = m, P = \mathbb{R}^n$). Condition [6.27] in this case is of the form

$$\hat{y}_0 f'(\hat{x}) + \sum_{i \in S(\hat{x})} \hat{y}_i g_i'(\hat{x}) = 0. \qquad [6.28]$$

For $\hat{y}_0 = 1$, this means that the anti-gradient of the objective function is a non-negative linear combination of the gradients of functions, which describe active constraints at the point $\hat{x}$.

PROOF.– Let us prove theorem 6.4. Consider the linear system relative to x:

$$\langle f'(\hat{x}), x - \hat{x}\rangle < 0, \qquad [6.29]$$

$$\langle g_i'(\hat{x}), x - \hat{x}\rangle < 0, \quad i \in I(\hat{x}), \qquad [6.30]$$

$$\langle g_i'(\hat{x}), x - \hat{x}\rangle = 0, \quad i = k+1, \ldots, m. \qquad [6.31]$$

Suppose that it has no solution on ri P. Then, by theorem 5.12 (Fan's theorem), there exist numbers $\hat{y}_0 \geq 0, \hat{y}_i \geq 0,\ i \in I(\hat{x}),\ \hat{y}_{k+1}, \ldots, \hat{y}_m$ not all equal to zero and such that

$$\langle \hat{y}_0 f'(\hat{x}), x - \hat{x}\rangle + \sum_{i \in S(\hat{x})} \hat{y}_i \langle g_i'(\hat{x}), x - \hat{x}\rangle =$$

$$= \langle \hat{y}_0 f'(\hat{x}) + \sum_{i \in S(\hat{x})} \hat{y}_i g_i'(\hat{x}), x - \hat{x}\rangle \geq 0$$

for all ri P. Since ri $P \subset \overline{P} = \overline{\text{ri}\, P}$, then this inequality holds for $x \in P$. In other words, condition [6.27] holds true. Hence the assertion of theorem 6.4 holds true.

Thus, in order to prove theorem 6.4, it remains to verify that the systems [6.29]–[6.31] have no solutions on ri P. We prove the statement of the theorem for the problem whose equality constraints are linear. In this case, it is easy to show that the systems [6.29]–[6.31] have no solution on P. Note that for the linear function $g(x) = \langle a, x\rangle + b$, the following formula holds:

$$g(x+h) = g(x) + \langle g'(x), h\rangle, \qquad [6.32]$$

where $g'(x) = a$. This formula allows us to substantially simplify the proof.

Let the systems [6.29]–[6.31] have a solution on P. Take $h = x - \hat{x}$. From inequality [6.29] and lemma 6.1, it follows that $h \in U(\hat{x}, f)$. For any $i \in I(\hat{x})$, from [6.30] and the same lemma 6.1, it follows that $h \in U(\hat{x}, g_i)$, that is $g_i(\hat{x} + \alpha h) < g_i(\hat{x}) = 0$ for all small enough $\alpha > 0$. For any $i \in \{1, \ldots, k\} \setminus I(\hat{x})$, we have $g_i(\hat{x}) < 0$ and $g_i(\hat{x} + \alpha h) < 0$ for all sufficiently small modulo α. For any $i = k+1, \ldots, m$, considering [6.31] and [6.29], we get

$$g_i(\hat{x} + \alpha h) = g_i(\hat{x}) + \langle g_i'(\hat{x}), h\rangle = g_i(\hat{x}) \qquad [6.33]$$

for all α. Finally, we note that from the convexity of P it follows that $\hat{x} + \alpha h = \alpha x + (1-\alpha)\hat{x} \in P \quad \forall \alpha \in [0, 1]$. Therefore, $\hat{x} + \alpha h \in X$ for all sufficiently small

$\alpha > 0$. In other words, $h \in U(\hat{x}, f) \cap V(\hat{x}, X)$. But this contradicts theorem 6.1. Consequently, the systems [6.29]–[6.31] have no solution on P. The remaining option is to use Fan's theorem. □

LEMMA 6.7.– (*Lyusternik's theorem*) Let the functions $g_1(x), \ldots, g_m(x)$ be continuously differentiable in some neighborhood of a point $\hat{x} \in \mathbb{R}^n$, and let gradients $g_1'(\hat{x}), \ldots, g_m'(\hat{x})$ together with some vectors $a_{m+1}, \ldots, a_n$ form a basis in $\mathbb{R}^n$. Assume that a vector $h \in \mathbb{R}^n$ satisfies the condition

$$\langle g_i'(\hat{x}), h \rangle = 0, \quad i = 1, \ldots, m.$$

Then there is an n-dimensional vector-valued function $r(\alpha), \alpha \in R$, such that

$$g_i\left(\hat{x} + \alpha h + r(\alpha)\right) = 0, \quad i = 1, \ldots, m,$$

$$\langle a_i, r(\alpha) \rangle = 0, \quad i = m+1, \ldots, n,$$

for all sufficiently small α and $\lim\limits_{\alpha \to 0} \frac{r(\alpha)}{\alpha} = 0$.

PROOF.– To prove lemma 6.7, we use the implicit function theorem.

THEOREM 6.6.– (*Implicit function theorem*) Let the functions $f_1(r, \alpha), \ldots, f_n(r, \alpha)$ be continuously differentiable in some neighborhood of zero in $\mathbb{R}^n \times \mathbb{R}$, let

$$f_i\,(0,\ 0) = 0, \quad i = 1, \ldots, n,$$

let gradients $\frac{\partial f_1}{\partial r}(0,0), \ldots, \frac{\partial f_n}{\partial r}(0,0)$ be linear independent and let $\frac{\partial f_i}{\partial \alpha}(0,0) = 0$, $i = 1, \ldots, n$. Then there exists such an n-dimension vector-function $r\,(\alpha)\,, \alpha \in R$, that

$$f_i\,(r\,(\alpha)\,, \alpha) = 0, \quad i = 1, \ldots, n,$$

for all sufficiently small α and $\lim\limits_{\alpha \to 0} \frac{r(\alpha)}{\alpha} = 0$.

Consider the functions

$$f_i\,(r, \alpha) = g_i\,(\hat{x} + \alpha\, h + r)\,, \quad i = 1, \ldots, m,$$

$$f_i\,(r, \alpha) = \langle a_i, r \rangle\,, \quad i = m+1, \ldots, n.$$

We have

$$\frac{\partial f_i}{\partial r}(0,\ 0) = g_i'\,(\hat{x})\,, \quad \frac{\partial f_i}{\partial \alpha}(0,\ 0) = \langle g_i'\,(\hat{x})\,, h \rangle\,, \quad i = 1, \ldots, m,$$

$$\frac{\partial f_i}{\partial r}(0,\ 0) = a_i\,, \quad \frac{\partial f_i}{\partial \alpha}(0,\ 0) = 0 \quad i = m+1, \ldots, n.$$

Here all conditions of the implicit function theorem are fulfilled. Using this theorem, we obtain the statement of lemma 6.7. □

PROOF.– Now prove theorem 6.4. Let $L = \operatorname{Lin} P$ be a parallel to P linear subspace. Consider any vectors $a_1, \ldots, a_s$ that form a basis in an orthogonal to L space $L^{\perp}$. Then L has the form

$$L = \{ h \in \mathbb{R}^n \mid \langle a_i, h \rangle = 0,\ i = 1, \ldots, s \} . \qquad [6.34]$$

Assume that vectors $g'_{k+1}(\hat{x}), \ldots, g'_m(\hat{x}),\ a_1, \ldots, a_s$ are linearly dependent, that is,

$$\sum_{i=k+1}^{m} \hat{y}_i g'_i(\hat{x}) + \sum_{j=1}^{s} \lambda_j a_j = 0 \qquad [6.35]$$

for some numbers $\hat{y}_{k+1}, \ldots, \hat{y}_m, \lambda_1, \ldots, \lambda_s$, not all equal to zero. Among the numbers $\hat{y}_{k+1}, \ldots, \hat{y}_m$ there are non-zero ones because of the linear independence of vectors a_1 , … , a_s. From [6.34] and [6.35] for any $h \in L$, we have

$$\left\langle \sum_{i=k+1}^{m} \hat{y}_i g'_i \left(\hat{x}\right), h \right\rangle = - \sum_{j=1}^{s} \lambda_j \left\langle a_j, h \right\rangle = 0.$$

But L can be written as $L = \operatorname{aff} P - \hat{x}$. That is why

$$\left\langle \sum_{i=k+1}^{m} \hat{y}_i g'_i \left(\hat{x}\right), \hat{x} - x \right\rangle = 0$$

for all $x \in \operatorname{aff} P$ and $x \in P$. Considering [6.21], the numbers $\hat{y}_{k+1}, \ldots, \hat{y}_m$ together with the numbers $\hat{y}_0 = \hat{y}_1 = \cdots = \hat{y}_k = 0$ satisfy the conditions [6.22] and [6.23], which proves the theorem in this degenerate case.

Now let the vectors $g'_{k+1}(\hat{x}), \ldots, g'_m(\hat{x}), a_1, \ldots, a_s$ be linear independent. We can assume that they form a basis in $\mathbb{R}^n$ (otherwise they are supplemented to the basis). Suppose that the systems [6.29]–[6.31] have a solution for $x \in \operatorname{ri} P$. Take $h = x - \hat{x}$. Considering [6.31] from lemma 6.7, we can conclude that there exists such an n-dimensional vector-function $r(\alpha)$, $\alpha \in R$, that

$$g_i \left(\hat{x} + \alpha\, h + r\left(\alpha\right)\right) = 0, \quad i = 1, \ldots, m, \qquad [6.36]$$

$$\left\langle a_i,\ r\left(\alpha\right)\right\rangle = 0, \quad i = m + 1, \ldots, n, \qquad [6.37]$$

for all sufficiently small α and

$$\lim_{\alpha \to 0} \frac{r\left(\alpha\right)}{\alpha} = 0. \qquad [6.38]$$

Take $x(\alpha) = \hat{x} + \alpha h + r(\alpha)$. Further, with an accuracy of $r(\alpha)$, the proof is reproduced for the case of linear equality constraints, as well as the proof of lemma 6.6 and theorem 6.1.)

From the differentiability of the function f at the point $\hat{x}$, it follows that

$$f(x(\alpha)) - f(\hat{x}) = \langle f'(\hat{x}), \alpha h + r(\alpha)\rangle + o(\alpha)$$

$$= \alpha \left(\langle f'(\hat{x}), h\rangle + \left\langle f'(\hat{x}), \frac{r(\alpha)}{\alpha} \right\rangle + \frac{o(\alpha)}{\alpha} \right).$$

Hence, taking into account [6.29] and [6.38], we get that $f(x(\alpha)) < f(\hat{x})$. For any $i \in I(\hat{x})$, making use of [6.30], we get that $g_i(x(\alpha)) < g_i(\hat{x}) = 0$ for all sufficiently small $\alpha > 0$. For any $i \in \{1, \ldots, k\} \backslash I(\hat{x})$, we have $g_i(\hat{x}) < 0$ and $g_i(x(\alpha)) < 0$ for all sufficiently small $\alpha > 0$. Formulas [6.34] and [6.37] mean that $r(\alpha) \in L = \operatorname{Lin} P$. Take $\bar{x}(\alpha) = x + \frac{r(\alpha)}{\alpha}$. Then $\bar{x} \in \operatorname{aff} P$. Since $x \in \operatorname{ri} P$ and $\bar{x} \to x$ by virtue of [6.38] for sufficiently small $\alpha > 0$, we have $\bar{x}(\alpha) \in P$ and

$$x(\alpha) = \hat{x} + \alpha(x - \hat{x}) + r(\alpha) = \alpha\,\bar{x}(\alpha) + (1 - \alpha)\,\hat{x} \in P.$$

Combining the above facts starting with [6.36], we get that $f(x(\alpha)) < f(\hat{x})$ and $x(\alpha) \in X$ for all sufficiently small $\alpha > 0$, and $x(\alpha) \to \hat{x}$. This contradicts the fact that $\hat{x}$ is a local solution to the problem [6.18]. Consequently, the systems [6.29]–[6.31] have no solutions on ri P. Using Fan's theorem, we complete the proof of the theorem. □

6.4.2. *Differential form of the Kuhn–Tucker theorem*

Recall that conditions of regularity are additional assumptions in the description of the problem [6.18] under which the equality $\hat{y}_0 = 1$ is guaranteed in theorem 6.4. The simplest example of these conditions is the requirement of linear independence of the gradients $g_1'(\hat{x}), \ldots, g_m'(\hat{x})$ in the classical constrained optimization problem. The condition of regularity in the problem [6.18] for $\hat{x} \in \operatorname{int} P$ is the linear independence of the gradients $g_i'(\hat{x}), i \in S(\hat{x})$. For a point $\hat{x} \in \operatorname{int} P$, formula [6.27] can be represented in the form

$$\hat{y}_0 f'(\hat{x}) + \sum_{i \in S(\hat{x})} \hat{y}_i g_i'(\hat{x}) = 0,$$

where numbers $\hat{y}_0, \hat{y}_i, i \in S(\hat{x})$, are not all equal to zero. The case $\hat{y}_0 = 0$ is impossible here because of the linear independence of $g_i'(\hat{x}), i \in S(\hat{x})$.

Unfortunately, the regularity conditions of this type are difficult to verify because they are formulated in terms of the point of minimum $\hat{x}$ that needs to be found. More convenient regularity conditions can be obtained for problems with convex constraints and linear equality constraints. The following theorem gives a group of such conditions.

THEOREM 6.7.– In the problem [6.18] let the set P be convex, the functions $f, g_1, \dots, g_k$ be differentiable at a point $\hat{x} \in X$, the functions $g_1, \dots, g_k$ be convex on P and the function $g_{k+1}, \dots, g_m$ be linear. Suppose that at least one of the following conditions is satisfied additionally:

1) there are no equality constraints ($k = m$) and there exists a point $\overline{x} \in P$ such that $g_i(\overline{x}) < 0,\ i = 1, \dots, m$;

2) the set P is a polyhedron and the functions $g_1, \dots, g_k$ are linear;

3) the set P is a polyhedron and the functions $g_{l+1}, \dots, g_k$, $0 < l \leq k$, are linear and there exists a point $\overline{x} \in X$ such that $g_i(\overline{x}) < 0 \ \ \forall i = 1, \dots, l$;

4) the functions $g_{l+1}, \dots, g_k$, $0 < l \leq k$, are linear and there exists a point $\overline{x} \in \operatorname{ri} P \cap X$ such that $g_i(\overline{x}) < 0 \ \forall i = 1, \dots, l$.

If $\hat{x}$ is a local solution of the problem [6.18], then there exists a vector $\hat{y} = (\hat{y}_1, \dots, \hat{y}_m) \in Q$ such that for $\hat{y}_0 = 1$ conditions [6.22] and [6.23] hold true.

PROOF.– Let condition (1) be satisfied. Then the systems [6.29] and [6.30] have no solution on P (relation [6.31] is absent here). At the same time for the specified point $\overline{x}$ and any $i \in I(\hat{x})$, using theorem 3.16, we obtain

$$\langle g_i'(\hat{x}), \overline{x} - \hat{x} \rangle \leq g_i(\overline{x}) - g_i(\hat{x}) = g_i(\overline{x}) < 0, \qquad [6.39]$$

that is $\overline{x}$ is a solution of the problem [6.30] on P. Then, by theorem 5.13 (Fan's theorem), there exist numbers $\hat{y}_i \geq 0$, $i \in I(\hat{x})$, such that for $\hat{y}_0 = 1$ condition [6.27] holds true with $S(\hat{x}) = I(\hat{x})$. This proves the theorem under condition (1).

Let condition (2) be satisfied. Consider the linear system

$$\langle f'(\hat{x}), x - \hat{x} \rangle < 0, \qquad [6.40]$$

$$\langle g_i'(\hat{x}), x - \hat{x} \rangle \leq 0, \quad i \in I(\hat{x}), \qquad [6.41]$$

$$\langle g_i'(\hat{x}), x - \hat{x} \rangle = 0, \quad i = k+1, \dots, m, \qquad [6.42]$$

which differs from [6.29]–[6.31] only with a non-strict sign in [6.41]. If x satisfies [6.41], then for $h = x - \hat{x}$ and any $i \in I(\hat{x})$, considering [6.29], we get

$$g_i(\hat{x} + \alpha h) = g_i(\hat{x}) + \alpha \langle g_i'(\hat{x}), h \rangle \leq g_i(\hat{x}) = 0$$

for all $\alpha \geq 0$. Next, we show that the systems [6.40]–[6.41] have no solution on P. At the same time, the point $\hat{x}$ itself is a solution of the systems [6.41] and [6.42]. Applying theorem 5.16, we again come to [6.27] for $\hat{y}_0 = 1$.

Let condition (3) be satisfied. Consider the linear system

$$\langle f'(\hat{x}), x - \hat{x}\rangle < 0, \qquad [6.43]$$

$$\langle g_i'(\hat{x}), x - \hat{x}\rangle < 0, \quad i \in I(\hat{x}) \cap \{1, ..., l\}, \qquad [6.44]$$

$$\langle g_i'(\hat{x}), x - \hat{x}\rangle \leq 0, \quad i \in I(\hat{x}) \cap \{l+1, \ldots, k\}, \qquad [6.45]$$

$$\langle g_i'(\hat{x}), x - \hat{x}\rangle = 0, \quad i = k+1, ..., m. \qquad [6.46]$$

This system has no solutions on P. At the same time, the point $\overline{x}$ is a solution of the systems [6.44]–[6.46] on P. Actually, for any $i \in I(\hat{x}) \cap \{1, \ldots, l\}$ condition [6.35] is satisfied. For any $i \in I(\hat{x}) \cap \{l+1, \ldots, k\}$, considering [6.29], we have

$$\langle g_i'(\hat{x}), \overline{x} - \hat{x}\rangle = g_i(\overline{x}) - g_i(\hat{x}) = g_i(\overline{x}) \leq 0,$$

and for any $i \in I(\hat{x}) \cap \{k+1, \ldots, m\}$ we can write the sequence with the sign of equality at the end. Now it is necessary to apply theorem 5.17. This again leads to [6.27] with $\hat{y}_0 = 1$.

Let condition (4) be satisfied. Then the systems [6.43]–[6.46] have no solution on P, wherein $\overline{x}$ is a solution of the system [6.44]–[6.46] on $\operatorname{ri} P$. Now it is necessary to apply theorem 5.18. □

From theorems 6.5 and 6.7 immediately follows one of the most important facts of the convex constrained optimization theory.

THEOREM 6.8.– (*Kuhn–Tucker theorem in differential form*) Let the conditions of theorem 6.7 be satisfied and let the function f be convex on P. A point $\hat{x}$ is a solution of the problem [6.18] if and only if there exists a vector $\hat{y} \in Q$ such that for $\hat{y}_0 = 1$ conditions [6.22] and [6.23] hold true.

REMARK 6.3.– Condition (1) in theorem 6.7 (and theorem 6.8) is called the *Slater condition*. This condition of regularity is simple and often used. Condition (2) is called the *linearity condition*. Note that it is automatically executed for linear and quadratic optimization problems. Conditions (3) and (4) are called *modified Slater conditions*. Their common point is their requirement of fulfillment of a Slater-type condition relative to nonlinear inequality constraints only. Their difference lies in the location of the point $\overline{x}$. In condition (3), this is just a point from the admissible set X. In condition (4), it is required, in addition, that $\overline{x}$ belong to the relative interior of the set of direct constraints P. Thus, the assumption "P is a polyhedron" and "$\overline{x} \in \operatorname{ri} P \cap X$" here replace each other. We note that conditions (3) and (4) for $P = \mathbb{R}^n$ merge into one.

6.4.3. *Second-order conditions of optimality*

Denote by

$$L''_{xx}(x, y_0, y) = y_0 f''(x) + \sum_{i=1}^{m} y_i g''_i(\hat{x})$$

a matrix composed of the second partial derivatives of the Lagrange function.

For a point $\hat{x} \in P$, we determine the set

$$V(\hat{x}) = \{h \in \mathbb{R}^n \,|h = \lambda(x - \hat{x}), \, \lambda > 0, \, x \in P\}.$$

It is clear that $V(\hat{x}) = \mathbb{R}^n$, if $\hat{x} \in \operatorname{int} P$.

Denote by $H(\hat{x})$ the set of all vectors $h \in \mathbb{R}^n$ such that

$$\langle f'(\hat{x}), h\rangle \leq 0, \qquad [6.47]$$

$$\langle g'_i(\hat{x}), h\rangle \leq 0, \quad i \in I(\hat{x}), \qquad [6.48]$$

$$\langle g'_i(\hat{x}), h\rangle = 0, \quad i = k+1, \ldots, m. \qquad [6.49]$$

We formulate a theorem about sufficient conditions for optimality in (not necessarily convex and regular) constrained optimization problems.

THEOREM 6.9.– In problem [6.18], let the functions $f, g_1, \ldots, g_m$ be two times differentiable at a point $\hat{x} \in X$. Assume that there exists a number $\hat{y}_0 \geq 0$ and a vector $\hat{y} \in Q$ such that the conditions [6.22] and [6.23] hold true and, moreover,

$$\langle L''_{xx}(\hat{x}, \hat{y}_0, \hat{y})h, h\rangle > 0 \qquad [6.50]$$

for all non-zero $h \in \overline{V(\hat{x})} \cap H(\hat{x})$. Then $\hat{x}$ is a strict local solution of the problem [6.18], that is $f(\hat{x}) < f(x)$ for all $x \in X$ close to $\hat{x}$, but different from $\hat{x}$.

PROOF.– If $\hat{x}$ is an isolated point of the set X, then this assertion is trivial. Let $\hat{x}$ be the boundary point of X, which is not a strict local solution of the problem [6.18]. Then there exists a sequence $\{x^k\}$ satisfying the conditions

$$x^k \in X, \, x^k \neq \hat{x}, \, x^k \to \hat{x}, \, f(x^k) \leq f(\hat{x}). \qquad [6.51]$$

Write x^k in the form

$$x^k = \hat{x} + \alpha_k h_k, \text{ where } \quad \alpha_k = \left\|x^k - \hat{x}\right\|, \quad h^k = (x^k - \hat{x})/\alpha_k.$$

Since $\|h^k\| = 1$, then we can assume that $h^k \to h \neq 0$. Since $h^k \in V(\hat{x})$, then $h \in \overline{V(\hat{x})}$. In this case, using [6.51], we have

$$0 \geq f(x^k) - f(\hat{x}) = \langle f'(\hat{x}), \alpha_k h^k \rangle + o(\alpha_k),$$
$$0 \geq g_i(x^k) - g_i(\hat{x}) = \langle g_i'(\hat{x}), \alpha_k h^k \rangle + o(\alpha_k), \ i \in I(\hat{x}),$$
$$0 = g_i(x^k) - g_i(\hat{x}) = \langle g_i'(\hat{x}), \alpha_k h^k \rangle + o(\alpha_k), \ i = k+1, \ldots, m$$

Dividing these relationships by α_k and passing to the limit, we obtain [6.47]–[6.49]. So $h \in \overline{V(\hat{x})} \cap H(\hat{x})$ and $h \neq 0$.

Next, from [6.22] it follows that

$$\langle L_x'(\hat{x}, \hat{y}_0, \hat{y}), h^k \rangle \geq 0. \tag{6.52}$$

Considering [6.23] and [6.51], we have

$$\begin{aligned} L(x^k, \hat{y}_0, \hat{y}) &= \hat{y}_0 f(x^k) + \sum_{i=1}^{m} \hat{y}_i g_i(x^k) \leq \hat{y}_0 f(x^k) \\ &\leq \hat{y}_0 f(\hat{x}) = \hat{y}_0 f(\hat{x}) + \sum_{i=1}^{m} \hat{y}_i g_i(\hat{x}) = L(\hat{x}, \hat{y}_0, \hat{y}). \end{aligned} \tag{6.53}$$

It follows from the conditions of the theorem that the function $L(x, \hat{y}_0, \hat{y})$ is twice differentiable at point $\hat{x}$. So,

$$L(x^k, \hat{y}_0, \hat{y}) = L(\hat{x}, \hat{y}_0, \hat{y}) + \langle L'(\hat{x}, \hat{y}_0, \hat{y}), \alpha_k h^k \rangle$$
$$+ \tfrac{1}{2} \langle L_{xx}''(\hat{x}, \hat{y}_0, \hat{y})(\alpha_k h^k), \alpha_k h^k \rangle + o(\alpha_k^2).$$

From this, and also from [6.52] and [6.53], we get

$$\frac{\alpha_k^2}{2} \langle L_{xx}''(\hat{x}, \hat{y}_0, \hat{y}) h^k, h^k \rangle + 0(\alpha_k^2) \leq 0.$$

Dividing both sides of this inequality by α_k^2 and passing to the limit, we obtain the inequality, which contradicts [6.50]. □

REMARK 6.4.– For any $\hat{y}_0 \geq 0$ and $\hat{y} \in Q$, which satisfy [6.22] and [6.23] under condition $h \in \overline{V(\hat{x})} \cap H(\hat{x})$, we have

$$\langle L_x'(\hat{x}, \hat{y}_0, \hat{y}), h \rangle = 0, \tag{6.54}$$

$$\hat{y}_0 \langle f'(\hat{x}), h \rangle = 0, \tag{6.55}$$

$$\hat{y}_i \langle g_i'(\hat{x}), h \rangle = 0, \quad i \in I(\hat{x}). \tag{6.56}$$

Actually, from [6.22] and the condition $h \in \overline{V(\hat{x})}$ it follows that $\langle L_x'(\hat{x}, \hat{y}_0, \hat{y}), h \rangle \geq 0$. From [6.23] and the condition $h \in H(\hat{x})$, we get the opposite inequality

$$\langle L_x'(\hat{x}, \hat{y}_0, \hat{y}), h \rangle = \langle \hat{y}_0 f'(\hat{x}) + \sum_{i \in S(\hat{x})} \hat{y}_i g_i'(\hat{x}), h \rangle$$

$$= \hat{y}_0 \langle f'(\hat{x}), h \rangle + \sum_{i \in S(\hat{x})} \hat{y}_i \langle g_i'(\hat{x}), h \rangle \leq 0.$$

This is possible only if the relations [6.54]–[6.56] hold true.

Theorem 6.9 allows some modifications. Consider one of them, which is sometimes more convenient to use.

COROLLARY 6.1.– In problem [6.18], let the functions $f, g_1, \dots, g_m$ be two times differentiable at a point $\hat{x} \in X$. Suppose that there exist points $\hat{y}_0 \geq 0$ and $\hat{y} \in Q$ such that conditions [6.22], [6.23] and [6.50] are satisfied for all non-zero $h \in \overline{V(\hat{x})}$, which satisfies [6.48], [6.49] and [6.56]. Then $\hat{x}$ is a local solution of the problem [6.18].

REMARK 6.5.– It is clear that for the classical constrained optimization problem this statement passes to theorem 1.18.

From theorem 6.9, one can also obtain a sufficient optimality condition using only first-order derivatives.

COROLLARY 6.2.– In problem [6.18], let the functions $f, g_1, \dots, g_m$ be differentiable at a point $\hat{x} \in X$. If $\overline{V(\hat{x})} \cap H(\hat{x}) = \{0\}$. Then $\hat{x}$ is a local solution of the problem [6.18].

REMARK 6.6.– For the convex constrained optimization problem, theorem 6.9 and its corollaries give sufficient conditions of uniqueness of the (global) solution of the problem.

We now give a theorem on the necessary second-order optimality condition, restricting ourselves to the case $\hat{x} \in \operatorname{int} P$.

THEOREM 6.10.– In problem [6.18], let the set P be convex, and let the functions $f, g_1, \dots, g_m$ be two times differentiable at a point $\hat{x} \in \operatorname{int} P \cap X$. Let, in addition, the functions $g_i'(\hat{x}), i \in S(\hat{x})$, be linear independent. If $\hat{x}$ is a strict local solution of the problem [6.18], then

$$\langle L''_{xx}(\hat{x}, \hat{y}_0, \hat{y})h, h \rangle \geq 0 \qquad [6.57]$$

for any $\hat{y}_0 \geq 0$ and $\hat{y} \in Q$, which satisfy conditions [6.23] and [6.25] and all $h \in H(\hat{x})$.

PROOF.– For arbitrary $h \in H(\hat{x})$, define sets of indices

$$I(\hat{x}, h) = \{i \in I(\hat{x}) | \langle g_i'(\hat{x}), h \rangle = 0\} = \{i | g_i(\hat{x}) = 0, \langle g_i'(\hat{x}), h \rangle = 0, \ 1 \leq i \leq k\},$$

$$S(\hat{x}, h) = I(\hat{x}, h) \cup \{k+1, \dots, m\} = \{i | g_i(\hat{x}) = 0, \ \langle g_i'(\hat{x}), h \rangle = 0, \ 1 \leq i \leq m\},$$

By the Lyusternick theorem, there exists an n-dimensional vector-valued function $r(\alpha)$, $\alpha \in \mathbb{R}$, such that

$$g_i(\hat{x} + \alpha h + r(\alpha)) = 0, \; i \in S(\hat{x}, h) \qquad [6.58]$$

for all sufficiently small α and

$$\lim_{\alpha \to 0} \frac{r(\alpha)}{\alpha} = 0. \qquad [6.59]$$

Take $x(\alpha) = \hat{x} + \alpha h + r(\alpha)$. For any $i \in \{1, \ldots, k\} \backslash I(\hat{x}, h)$, one of two conditions is satisfied:

$$g_i(\hat{x}) < 0, \quad \text{or} \quad \langle g_i'(\hat{x}), h \rangle < 0. \qquad [6.60]$$

Referring in the first case to [6.23], and in the second case on [6.39], we have $\hat{y}_i = 0$. From this and [6.58], it follows that $\hat{y}_i g_i(x(\alpha)) = 0$ for all $i = 1, \ldots, m$ and all sufficiently small $\alpha > 0$. Then

$$L(x(\alpha), \hat{y}_0, \hat{y}) = \hat{y}_0 f(x(\alpha)) + \sum_{i=1}^{m} \hat{y}_i g_i(x(\alpha)) = \hat{y}_0 f(x(\alpha)) \qquad [6.61]$$

for sufficiently small α.

Then from [6.58]–[6.60] and condition $\hat{x} \in \operatorname{int} P$, it is easy to deduce that $x(\alpha) \in X$ for all sufficiently small $\alpha > 0$, since $\hat{x}$ is a local solution of the problem [6.18]. Then, using formulas [6.23], [6.25] and [6.42], and taking into account the twice differentiability of the function $L(x, \hat{y}_0, \hat{y})$ at $\hat{x}$, we can write:

$$0 \leq \hat{y}_0 f(x(\alpha)) - \hat{y}_0 f(\hat{x}) = L(x(\alpha), \hat{y}_0, \hat{y}) - L(\hat{x}, \hat{y}_0, \hat{y})$$

$$= \frac{1}{2} \langle L''_{xx}(\hat{x}, \hat{y}_0, \hat{y}) h(\alpha), h(\alpha) \rangle + o(\alpha^2),$$

where $h(\alpha) = x(\alpha) - \hat{x} = \alpha h + r(\alpha)$. Hence

$$\frac{1}{2} \left\langle L''_{xx}(\hat{x}, \hat{y}_0, \hat{y}) \left(h + \frac{r(\alpha)}{\alpha} \right), \left(h + \frac{r(\alpha)}{\alpha} \right) \right\rangle + \frac{o(\alpha^2)}{\alpha^2} \geq 0.$$

Passing to the limit, we obtain [6.57]. □

REMARK 6.7.– For the classical constrained optimization problem, assertions of theorems 6.10 and 1.17 are similar.

6.5. Exercises

1) Give an example of a problem for which both factors in condition [6.23] are equal to zero.

2) Show that in theorem 6.4 the Lagrange multipliers $\hat{y}_0, \hat{y}_1, \dots, \hat{y}_m$ can be chosen in such a way that no more than $n + 1$ of them will be non-zero.

3) Ensure that in theorem 6.5 the functions g_i, $k + 1 \le i \le m$, can be considered convex if $\hat{y}_i \ge 0$, and can be considered concave if $\hat{y}_i \le 0$.

4) Show that in condition (2) of theorem 6.7 the functions $g_1, \dots, g_k$ can be considered concave on P. Hint: ensure that in this case the systems [6.40]–[6.42] have no solution on P.

5) In problem [6.18], let the set P be convex, the functions $f, g_1, \dots, g_k$ be differentiable at a point $\hat{x} \in X$ and let the functions $g_{k+1}, \dots, g_m$ be linear. Assume that at least one of the following conditions is satisfied additionally:

a) there are no equality constraints ($k = m$), and the system [6.30] has solutions on P;

b) the set P is a polyhedron, the functions $g_{l+1}, \dots, g_m (0 < l \le k)$ are linear and the systems [6.44]–[6.46] have solutions on P;

c) the functions $g_{l+1}, \dots, g_m$ $(0 \le l \le k)$ are linear, and the systems [6.44]–[6.46] have solutions on $\operatorname{ri} P$.

Prove that in theorem 6.4, we can take $\hat{y}_0 = 1$.

6) Show that if the conditions of theorem 6.10 are satisfied with $\hat{y}_0 = 0$, then $\hat{x}$ is an isolated point of the set X. Check that this is the case in the problem

$$x_1 \to \min,\ x_1^2 + x_2^2 \le 1,\ x_1^3 + x_2^3 = 1.$$

7) Solving the problem

$$x_1 \to \min,\ x_1^2 + x_2^2 \ge 1,\ x_1^3 + x_2^3 = 1,$$

make sure that the condition of linear independence of the gradients $g_i'(\hat{x}),\ i \in S(\hat{x})$ is essential in theorem 6.10.

8) Find solutions to the following constrained optimization problems:

a) $x_1^2 + (x_2 - 1)^2 \to \min, \quad x_1^2 + 4x_2^2 \le 4,\ 2x_1^2 + x_2 \ge 2,\ x_1 \ge 2x_2;$

b) $x_1 \to \max,\ x_1^2 + x_2^2 \le 1,\ (x_1 - 1)^2 + x_2^2 \ge 1,\ x_1 + x_2 \le 1;$

c) $10(x_1 - 3,5)^2 + 20(x_2 - 4)^2 \to \min,\ x_1 + x_2 \le 6,\ x_1 - x_2 \le 1,\ 2x_1 + x_2 \ge 6,\ 0,5x_1 - x_2 \ge -4,\ x_1 \ge 1,\ x_2 \ge 1;$

d) $25(x_1-2)^2+(x_2-2)^2 \to \max,\ x_1+x_2 \geq 2,\ x_1-x_2 \geq -2,\ x_1+x_2 \leq 6,\ x_1-3x_2 \leq 2,\ x_1 \geq 0,\ x_2 \geq 0.$

9) Suggest a method for solving the problem

$$\sum_{j=1}^{n}(x_j-a_j)^2 \to \min, \quad \sum_{j=1}^{n} x_j \leq 1, \quad x_1 \geq 0, \ldots, x_n \geq 0.$$

10) Solve the problems:

a) $\sum\limits_{j=1}^{n} \alpha_j\sqrt{x_j} \to \max,\ \ \sum\limits_{j=1}^{n} x_j \leq 1,\ \ x_i \geq 0,\ \alpha_i > 0,\ i = 1\ldots,n;$

b) $\prod\limits_{j=1}^{n} x_j^{\lambda_j} \to \max,\ \ \sum\limits_{j=1}^{n} p_j x_j \leq 1,\ x_i \geq 0,\ \lambda_i > 0,\ p_i > 0,\ i = 1\ldots,n;$

c) $\sum\limits_{j=1}^{n} (x_j-a_j)^2 \to \min,\ \ \sum\limits_{j=1}^{n} x_j^2 \leq 1,\ \ \sum\limits_{j=1}^{n} x_j = 0.$

11) Solve the problem:

$$ax_1^2+bx_1x_2+cx_2^2 \to \min, \quad x_1^2+x_2^2 \leq 1, \quad x_1+x_2 \geq 1,$$

for all possible values of a, b and c. Pay attention to how the assumption of convexity of the objective function: $a \geq 0, c \geq 0, 4ac \geq b^2$ simplifies the problem.

12) Ensure that $(-1,-1)$ is a stationary point of the problem

$$\frac{1}{3}x_1^3+x_2 \to \min, \quad x_1^2+x_2^2 \leq 2,$$

which falls into the "gap" between theorems 6.9 and 6.10: condition [6.40] is satisfied here as the equality for all h satisfying [6.47]–[6.49]. Find out if this point is a solution to the problem.

6.6. Dual problems in convex optimization

Consider the optimization problem

$$\begin{aligned}
&f(x) \to \min,\\
&g_i(x) \leq 0, \quad i = 1,\ldots,k;\\
&g_i(x) = 0, \quad i = k+1,\ldots,m;\\
&x \in P \subset \mathbb{R}^n.
\end{aligned} \qquad [6.62]$$

Denote by

$$X = \{x \in P\,|\,g_i(x) \leq 0,\ i = 1,\ldots,k;\ \ g_i(x) = 0,\ i = k+1,\ldots,m\}$$

the admissible set of the problem [6.62]. Denote by

$$Q = \{y = (y_1, \ldots, y_m) \in \mathbb{R}^m \mid y_i \geq 0, i = 1, \ldots, k\}$$

the set of vectors from $\mathbb{R}^m$ in which the first k coordinates are non-negative. Let

$$L(x, y) = f(x) + \sum_{i=1}^{m} y_i g_i(x)$$

be the regular Lagrange function of the problem [6.62].

Assume that $X \neq \emptyset$. Denote by $\hat{f}$ the exact lower bound of the objective function of the problem [6.62] on its admissible set:

$$\hat{f} = \inf_{x \in X} f(x).$$

We will call $\hat{f}$ the value of the problem [6.62]. It is clear that the point $\hat{x} \in X$ is a (global) solution of the problem [6.62] only if $f(\hat{x}) = \hat{f}$. However, it can also happen that the problem [6.62] has no solution, that is, $f(x) > \hat{f} \geq -\infty$ for all $x \in X$.

6.6.1. *Kuhn–Tucker vector*

DEFINITION 6.3.– *A vector $y \in Q$ is called the Kuhn–Tucker vector of problem [6.62], if*

$$\hat{f} \leq f(x) + \sum_{i=1}^{m} y_i g_i(x) = L(x, y) \quad \textit{for all} \quad x \in P. \qquad [6.63]$$

Problems for which such a vector exists have a number of properties that are absent in more general cases. It turns out that the Kuhn–Tucker vector exists for a sufficiently large class of convex optimization problems. Before we prove the corresponding result, we will prove a weaker statement, which reflects one of the characteristic properties of convex optimization problems.

THEOREM 6.11.– In problem [6.62], let the set P be convex, the functions $f(x), g_1(x), \ldots, g_k(x)$ be convex on P, the functions $g_{k+1}(x), \ldots, g_m(x)$ be linear and let the set X be non-empty. Then there exist a number $\hat{y}_0 \geq 0$ and a vector $\hat{y} \in Q$, which are not all equal to zero and such that

$$\hat{y}_0 \hat{f} \leq \hat{y}_0 f(x) + \sum_{i=1}^{m} \hat{y}_i g_i(x) = L(x, \hat{y}_0, \hat{y}) \quad \text{for all} \quad x \in P. \qquad [6.64]$$

PROOF.– If $\hat{f} = -\infty$, then [6.64] holds true for all $\hat{y}_0 \geq 0$ and $\hat{y} \in Q$. Let $\hat{f} > -\infty$. Consider the system

$$\begin{aligned} & f(x) - \hat{f} < 0; \\ & g_i(x) < 0, \quad i = 1, \dots, k; \\ & g_i(x) = 0, \quad i = k+1, \dots, m. \end{aligned}$$

By definition of $\hat{f}$, this system has no solutions on P. Then, by theorem 5.12 (Fan's theorem) there exists a number $\hat{y}_0 > 0$ and a vector $\hat{y} \in Q$ that are not all equal to zero and such that

$$\hat{y}_0(f(x) - \hat{f}) + \sum_{i=1}^{m} \hat{y}_i g_i(x) \geq 0 \quad \text{for all} \quad x \in P. \qquad [6.65]$$

This is relation [6.64]. □

REMARK 6.8.– Theorem 6.11 cannot be applied to all optimization problems. For example, it cannot be applied to the problem

$$-x^2 \to \min, \quad x = 0, \quad x \in P = R,$$

nor to the problem

$$x - 1 \to \min, \quad x^2 - 1 = 0, \quad x \in P = R_+.$$

In the first problem, the objective function is not convex, and in the second problem the equality constraint is nonlinear.

As we see, theorem 6.11 is a consequence of Fan's theorem, and the case $\hat{y}_0 = 0$ is not excluded in [6.64]. Using alternative results (regularity theorems) instead of Fan's theorem, we can specify additional conditions for which the case $\hat{y}_0 = 1$ is guaranteed in [6.64], that is, the existence of the Kuhn–Tucker vector is guaranteed.

THEOREM 6.12.– In problem [6.62], let the set P be convex, the functions $f(x)$, $g_1(x), \dots, g_k(x)$ be convex on P and let the functions $g_{k+1}(x), \dots, g_m(x)$ be linear. Suppose that at least one of the following conditions is satisfied additionally:

1) there are no equality constraints ($k = m$) and there exists a point $\overline{x} \in P$ such that $g_i(\overline{x}) < 0$ for all $i = 1, \dots, m$;

2) the set P is a polyhedron and the functions $g_1(x), \dots, g_k(x)$ are linear, the set X is non-empty;

3) the set P is a polyhedron and the functions $f(x), g_1(x), \dots, g_l(x)$, $0 \leq l \leq k$, are convex on a relatively open convex set U, which contains P, the functions $g_{l+1}(x), \dots, g_k(x)$ are linear and there exists a point $\overline{x} \in X$ such that $g_i(\overline{x}) < 0$ for all $i = 1, \dots, l$;

4) the functions $g_{l+1}(x), \dots, g_k(x)$, $0 < l \le k$, are linear and there exists a point $\overline{x} \in \operatorname{ri} P \cap X$ such that $g_i(\overline{x}) < 0$ for all $i = 1, \dots, l$.

Then the Kuhn–Tucker vector of the problem [6.62] exists.

PROOF.– The case $\hat{f} = -\infty$ is trivial. Let $\hat{f} > -\infty$. If condition (1) is fulfilled, then the system

$$\begin{aligned} &f(x) - \hat{f} < 0; \\ &g_i(x) < 0, \quad i = 1, \dots, m, \end{aligned}$$

which is considered on the set P satisfies the assumptions of theorem 5.13. Therefore, there exists a vector $y \in R_+^m = Q$ such that relation [6.65] holds for $\hat{y}_0 = 1$, that is, relation [6.63].

If condition (2) is satisfied, then the existence of the Kuhn–Tucker vector follows from theorem 5.16 applied to the system

$$\begin{aligned} &f(x) - \hat{f} < 0; \\ &g_i(x) \le 0, \quad i = 1, \dots, k; \\ &g_i(x) = 0, \quad i = k+1, \dots, m \end{aligned}$$

on the set P.

If condition (3) or (4) is satisfied, then we need to apply theorem 5.17 or 5.18 accordingly to the system

$$\begin{aligned} &f(x) - \hat{f} < 0; \\ &g_i(x) < 0, \quad i = 1, \dots, l; \\ &g_i(x) \le 0, \quad i = l+1, \dots, k, \\ &g_i(x) = 0, \quad i = k+1, \dots, m. \end{aligned}$$

on the set P. □

REMARK 6.9.– Conditions (1) and (4) of theorem 6.12, respectively, coincide with conditions (1) and (4) of theorem 6.7. Recall that condition (1) is called the Slater condition, and condition (4) is called the modified Slater condition. In conditions (2) and (3) of theorem 6.12, the requirement is a little more than in conditions (2) and (3) of theorem 6.7. This is essential. For example, for the convex optimization problem of the form

$$f(x) = -\sqrt{x_1 x_2} \to \min, \quad x_1 \le 0, \quad x \in P = R_+^2, \qquad [6.66]$$

the relation [6.64] holds true only when $\hat{y}_0 = 0$. None of the conditions (1)–(4) is satisfied; moreover, condition (2) is not satisfied because the function f is nonlinear, and condition (3) is not satisfied because the function f is convex only on P. Note that for this problem we cannot apply theorem 6.7, since at point $\hat{x} = 0$, which is a solution, the function f is not differentiable.

6.6.2. *Dual optimization problems*

Each optimization problem can be associated with the so-called dual (conjugate) optimization problem.

DEFINITION 6.4.– *The dual problem of problem [6.62] is*

$$\varphi(y) \to \max, \quad y \in Y, \tag{6.67}$$

where

$$\varphi(y) = \inf_{x \in P} L(x, y) = \inf_{x \in P} \left(f(x) + \sum_{i=1}^{m} y_i g_i(x) \right),$$

$$Y = \{y \in Q | \varphi(y) > -\infty\}.$$

In this case, problem [6.62] is called the direct problem. Assuming that $Y \neq \emptyset$, we denote by

$$\hat{\varphi} = \sup_{y \in Y} \varphi(y)$$

the value of problem [6.67].

REMARK 6.10.– Dual problem [6.67] can be written in the form

$$\varphi(y) \to \max, \quad y \in Q,$$

thus admitting infinite values of the function $\varphi(y)$. At the same time, the direct problem [6.62] can be written in the form

$$\psi(x) \to \min, \quad x \in P,$$

where

$$\psi(x) = \sup_{y \in Q} L(x, y) = \begin{cases} f(x), & \text{if} \quad x \in X \\ +\infty, & \text{if} \quad x \in P \backslash X. \end{cases}$$

We assume that $\hat{f} = +\infty$, if $X = \emptyset$, that is $\sup\limits_{y \in Q} L(x, y) = +\infty$ for all $x \in P$; $\hat{\varphi} = -\infty$, if $Y = \emptyset$, that is $\inf\limits_{x \in P} L(x, y) = -\infty$ for all $y \in Q$. Then we can write:

$$\hat{f} = \inf_{x \in P} \sup_{y \in Q} L(x, y), \quad \hat{\varphi} = \sup_{y \in Q} \inf_{x \in P} L(x, y). \tag{6.68}$$

Thus, the direct problem and the dual problem are determined symmetrically with respect to the Lagrange function $L(x, y)$ of the direct problem. To obtain the dual problem, it is enough to rearrange the operations $\inf_x$ and $\sup_y$ over this function.

We will show that the dual problem for each optimization problem is always a convex optimization problem if we consider it as a minimization problem.

THEOREM 6.13.– In problem [6.67], the set Y is convex and the function φ is concave on Y.

PROOF.– The Lagrange function $L(x, y)$ is linear with respect to y for all $x \in P$. Considering this for any $y^1, y^2 \in Q$ and $\lambda \in [0, 1]$, we have

$$\varphi(\lambda y^1 + (1-\lambda)y^2) = \inf_{x \in P} L(x, \lambda y^1 + (1-\lambda)y^2)$$

$$= \inf_{x \in P} (\lambda L(x, y^1) + (1-\lambda)L(x, y^2))$$

$$\geq \lambda \inf_{x \in P} L(x, y^1) + (1-\lambda) \inf_{x \in P} L(x, y^2) = \lambda\varphi(y^1) + (1-\lambda)\varphi(y^2).$$

It follows from this that the set Y is convex (from condition $\varphi(y^1) > -\infty$, $\varphi(y^2) > -\infty$ we have $\varphi(\lambda y^1 + (1-\lambda)y^2) > -\infty$, and the function φ is concave on Y). □

The following theorem establishes a relationship between the optimization problem and its dual problem.

THEOREM 6.14.– 1) For arbitrary $x \in X$, $y \in Q$, the inequality

$$f(x) \geq \varphi(y) \qquad [6.69]$$

holds true.

2) If $X \neq \emptyset$, $Y \neq \emptyset$, then

$$\hat{f} \geq \hat{\varphi}, \qquad [6.70]$$

that is, the value of the direct problem (minimization problem) is never less than the value of the dual problem (maximization problem).

PROOF.– For arbitrary $x \in X$, $y \in Q$, the inequality

$$f(x) \geq f(x) + \sum_{i=1}^{m} y_i g_i(x) = L(x, y) \geq \inf_{\hat{x} \in P} L(\hat{x}, y) = \varphi(y),$$

holds true, that is, relation [6.69] is satisfied. Hence, we obtain [6.70]. □

In inequality [6.70], the case $\hat{f} > \hat{\varphi}$ is possible. However, the central problem of the theory of duality is the search for conditions in which the values of direct and dual problems coincide, that is, $\hat{f} = \hat{\varphi}$, or, taking into account [6.68],

$$\inf_{x \in P} \sup_{y \in Q} L(x, y) = \sup_{y \in Q} \inf_{x \in P} L(x, y).$$

From this equality, which is called the *duality relation*, there follows a number of important consequences. In particular, this relationship allows us to replace the search for solutions for a direct problem with solutions for a dual problem, which is sometimes simpler.

We formulate the main result of the theory of duality.

THEOREM 6.15.– (*Duality theorem*) Suppose that the assumptions of theorem 6.12 are satisfied. If the value of the direct problem [6.62] is finite ($\hat{f} > -\infty$), then the set of solutions of the dual problem [6.67] is non-empty and coincides with the set of Kuhn–Tucker vectors of the problem [6.62]. At the same time, the duality relation

$$\hat{f} = \hat{\varphi} \qquad [6.71]$$

holds true.

PROOF.– Let $\hat{y} \in Q$ be the Kuhn–Tucker vector of the problem [6.62], which exists due to theorem 6.12. Then from [6.63], it follows that

$$\hat{f} \leq \inf_{x \in P} L(x, \hat{y}) = \varphi(\hat{y}) \leq \hat{\varphi}. \qquad [6.72]$$

Since $\hat{f} > -\infty$, then $\varphi(\hat{y}) > -\infty$, that is $\hat{y} \in Y$. Combining [6.70] and [6.72], we get [6.71], wherein $\varphi(\hat{y}) = \hat{\varphi}$, that is $\hat{y}$ is a solution of the problem [6.67]. So the arbitrary Kuhn–Tucker vector of the problem [6.62] is a solution of the problem [6.67].

Suppose, on the contrary, that $\hat{y}$ is a solution of the problem [6.67]. Then, using relation [6.71], we have

$$\hat{f} = \hat{\varphi} = \varphi(\hat{y}) = \inf_{x \in P} L(x, \hat{y}).$$

That is, relation [6.63] holds true. This means that y is the Kuhn–Tucker vector of the problem [6.67]. □

Another important statement follows from theorem 6.15.

THEOREM 6.16.– Suppose that the assumptions of theorem 6.12 are fulfilled. If the admissible set Y of the dual problem [6.67] is non-empty, then it has a solution. If $Y = \emptyset$, then the value of the direct problem [6.62] is infinite ($\hat{f} = -\infty$).

PROOF.– If $Y \neq \emptyset$, then $\hat{f} \geq \hat{\varphi} > -\infty$ due to theorem 6.14. Then by theorem 6.15 the problem [6.67] has solutions. If $Y = \emptyset$, then by theorem 6.15 the case $\hat{f} > -\infty$ is impossible. □

We give below another theorem on the connection between the optimization problem and its dual problem.

THEOREM 6.17.– In problem [6.62], let the set P be closed and convex, let the functions $f(x), g_1(x), \dots, g_k(x)$ be continuous and convex on P, let the functions $g_{k+1}(x), \dots, g_m(x)$ be linear and let the set of solutions of this problem be non-empty and bounded. Then $Y \neq \emptyset$ and $\hat{f} = \hat{\varphi}$.

PROOF.– We restrict ourselves to the fact that in problem [6.62] there are no equality constraints ($k = m$). Consider for any $\varepsilon > 0$ the problem

$$\begin{aligned} & f(x) \to \min, \\ & g_i(x) \le \varepsilon, \ i = 1, \dots, m, \\ & x \in P \subset \mathbb{R}^n. \end{aligned} \qquad [6.73]$$

For $\varepsilon = 0$, this is problem [6.62]. Let X_ε be the admissible set of problem [6.73], and let $\hat{f}_\varepsilon = \inf\limits_{x \in X_\varepsilon} f(x)$ be its value. Since $X \subset X_\varepsilon$, then $\hat{f}_\varepsilon \le \hat{f}$.

From theorem 3.28, it follows that problem [6.73] has as a solution $\hat{x}_\varepsilon \in X_\varepsilon$ and $f(\hat{x}_\varepsilon) = \hat{f}_\varepsilon$. For arbitrary $\varepsilon \in (0, 1]$, the point $\hat{x}_\varepsilon$ satisfies the system

$$\begin{aligned} & f(x) \le \hat{f}, \\ & g_i(x) \le 1, \ i = 1, \dots, m. \end{aligned} \qquad [6.74]$$

But, as a consequence of theorem 3.27, the set of solutions of this system is bounded (if in [6.74] we substitute 1 with 0, then we obtain a system that determines the set of solutions of the problem [6.62], which, under the conditions of the theorem, is not empty and bounded). Therefore, we can assume that $\hat{x}_\varepsilon \to \hat{x}$ as $\varepsilon \to 0$. Therefore $\hat{x} \in X$. Since $f(\hat{x}_\varepsilon) \le \hat{f}$ we have $f(\hat{x}) \le \hat{f}$. Hence, by the definition of $\hat{f}$, we have $f(\hat{x}) = \hat{f}$, that is, $\hat{x}$ is a solution of the problem [6.62]. Therefore,

$$\lim_{\varepsilon \to 0} \hat{f}_\varepsilon = \lim_{\varepsilon \to 0} f(\hat{x}_\varepsilon) = f(\hat{x}) = \hat{f}. \qquad [6.75]$$

Consider now the dual problem of problem [6.73]:

$$\varphi_\varepsilon(y) \to \max, \quad y \in Y_\varepsilon, \qquad [6.76]$$

where

$$\varphi_\varepsilon(y) = \inf_{x \in P} \left(f(x) + \sum_{i=1}^{m} y_i \left(g_i(x) - \varepsilon \right) \right), \qquad Y_\varepsilon = \{ y \in Q \,|\, \varphi_\varepsilon(y) > -\infty \}.$$

For arbitrary $y \in Q$, we have $\varphi_\varepsilon(y) \le \varphi(y)$. Hence $Y_\varepsilon \subset Y$ and

$$\hat{\varphi}_\varepsilon = \sup_{y \in Y_\varepsilon} \varphi_\varepsilon(y) \le \sup_{y \in Y} \varphi(y) = \hat{\varphi}. \qquad [6.77]$$

Problem [6.73] satisfies condition (1) of theorem 6.12 (any point $\bar{x} \in X$ is suitable). From theorem 6.15, it follows that, first, problem [6.76] has a solution. That is why $Y_\varepsilon \neq \emptyset$. Hence $Y \neq \emptyset$, since $Y_\varepsilon \subset Y$. And, second, $\hat{f}_\varepsilon = \hat{\varphi}_\varepsilon$. From here, considering [6.75] and [6.77], we get $\hat{f} \leq \hat{\varphi}$. Then $\hat{f} = \hat{\varphi}$ by virtue of [6.70]. □

We will point out one sufficient condition under which the set of solutions to problem [6.62] is non-empty.

THEOREM 6.18.– In problem [6.62], let the set P be closed and convex, let the functions $f(x), g_1(x), \ldots, g_k(x)$ be continuous and convex on P, let the functions $g_{k+1}(x), \ldots, g_m(x)$ be linear and let the set X be non-empty. Assume that for some $y \in Y$, the set $P(y)$ of all points from P such that $\varphi(y) = \inf_{x \in P} L(x, y)$, that is

$$P(y) = \left\{ x^0 \in P \,\middle|\, L\left(x^0, y\right) = \min_{x \in P} L\left(x, y\right) \right\},$$

is non-empty and bounded. Then the set of solutions to the problem [6.62] is non-empty and bounded.

PROOF.– Fix an arbitrary point $x^0 \in P\left(y\right)$ and consider the problem

$$\begin{aligned} & f(x) \to \min, \\ & g_i(x) \leq g_i(x^0), \quad i = 1, \ldots, k, \\ & g_i(x) = g_i(x^0), \quad i = k+1, \ldots, m, \\ & x \in P \subset \mathbb{R}^n. \end{aligned} \qquad [6.78]$$

We first show that x^0 is a solution of this problem. Let x be an admissible point of the problem. By choice of x^0, we have $L\left(x^0, y\right) \leq L\left(x, y\right)$, that is

$$f(x^0) + \sum_{i=1}^{m} y_i g_i(x^0) \leq f(x) + \sum_{i=1}^{m} y_i g_i(x).$$

Hence

$$f(x^0) \leq f(x) + \sum_{i=1}^{m} y_i \left(g_i(x) - g_i(x^0)\right) \leq f(x),$$

that is x^0 is a solution of the problem [6.78]. Let now $\bar{x}$ be an arbitrary solution of problem [6.78]. Then $f\left(\bar{x}\right) = f\left(x^0\right)$ and

$$L\left(\bar{x}, y\right) = f\left(\bar{x}\right) + \sum_{i=1}^{m} y_i g_i\left(\bar{x}\right) \leq$$

$$\leq f\left(x^0\right) + \sum_{i=1}^{m} y_i g_i\left(x^0\right) = L\left(x^0, y\right) = \min_{x \in P} L\left(x, y\right),$$

that is, $\bar{x} \in P(y)$. Consequently, the set of solutions of problem [6.78] is non-empty and, as a subset of the set $P(y)$, it is bounded. Then, by theorem 3.28, the set of solutions to problem [6.62] is also non-empty and bounded. □

6.6.3. *Kuhn–Tucker theorem for non-differentiable functions*

In the previous sections, we proved theorems establishing the necessary and sufficient conditions of optimality in the convex optimization problems for differentiable functions. Below are theorems that do not use derivatives and do not require differentiability of functions.

THEOREM 6.19.– *(Kuhn–Tucker theorem in the form of duality)* Suppose that the assumptions of theorem 6.12 are fulfilled. The point $\hat{x} \in X$ is a solution of problem [6.62] if and only if there exists a vector $\hat{y} \in Q$ such that the duality ratio holds true

$$f(\hat{x}) = \varphi(\hat{y}), \qquad [6.79]$$

which is equivalent to the conditions

$$L(\hat{x}, \hat{y}) \leq \min_{x \in P} L(x, \hat{y}), \qquad [6.80]$$

$$\hat{y}_i g_i(\hat{x}) = 0, \quad i = 1, \ldots, k. \qquad [6.81]$$

The set of vectors $\hat{y} \in Q$ satisfying [6.79] coincides with the set of solutions of the dual problem [6.67] or (see theorem 6.15) with the set of Kuhn–Tucker vectors of the direct problem [6.62].

PROOF.– The most significant statement of the theorem was actually proved earlier: if $\hat{x}$ is a solution of problem [6.62], then by theorem 6.15 problem [6.67] has solutions and any of its solutions $\hat{y}$ satisfies [6.79] because

$$f(\hat{x}) = \hat{f} = \hat{\varphi} = \varphi(\hat{y}).$$

Further reasonings are valid for any problem of the form [6.62].

Let relation [6.79] be satisfied. Then for any $x \in X$, using [6.69], we obtain

$$f(\hat{x}) = \varphi(\hat{y}) \leq f(x),$$

that is $\hat{x}$ is a solution of problem [6.62]. Similarly, for any $y \in Y$ we have

$$\varphi(\hat{y}) = f(\hat{x}) \geq \varphi(y),$$

that is, $\hat{y}$ is a solution of problem [6.67]. It remains to show that equality [6.79] is equivalent to conditions [6.80] and [6.81]. Let relation [6.79] be satisfied. Then by definition of φ, we have

$$f(\hat{x}) = \varphi(\hat{y}) \leq f(x) + \sum_{i=1}^{m} \hat{y}_i g_i(x) \qquad [6.82]$$

for all $x \in P$. Substituting here $x = \hat{x}$, we obtain

$$\sum_{i=1}^{m} \hat{y}_i g_i(\hat{x}) \geq 0.$$

But $\hat{x} \in X$, that is

$$g_i(\hat{x}) \leq 0, \quad i = 1, \ldots, k, g_i(\hat{x}) = 0, \quad i = k+1, \ldots, m.$$

Hence

$$\sum_{i=1}^{m} \hat{y}_i g_i(\hat{x}) = 0 \qquad [6.83]$$

and relation [6.81] holds true. From [6.83], taking into account the definition of L, it follows that

$$L(\hat{x}, \hat{y}) = f(\hat{x}). \qquad [6.84]$$

Therefore, [6.82] can be rewritten in the form

$$L(\hat{x}, \hat{y}) \leq L(x, \hat{y}) \text{ for all } \quad x \in P.$$

And this is the same as [6.80].

On the contrary, let the conditions [6.80] and [6.81] be satisfied. From relation [6.81], the relation [6.83] follows and therefore [6.84] holds true. Then the relation [6.80] is of the form [6.79]. □

REMARK 6.11.– In the case where the functions $f, g_1, \ldots, g_k$ are differentiable at the point $\hat{x}$, condition [6.80] is equivalent to condition [6.22] with $\hat{y}_0 = 1$ (theorem 6.2). At the same time, condition [6.36] simply coincides with condition [6.23]. Consequently, theorem 6.19 is a generalization of theorem 6.8 to the case of non-differentiable functions. In relation to this it is important to emphasize that the concept of the Kuhn–Tucker vector is a generalization of the concept of the Lagrange multiplier vector (that is, the vector $\hat{y} \in Q$ satisfying the conditions [6.22] and [6.23] with $\hat{y}_0 = 1$). As mentioned previously, it is clear that within the conditions of theorem 6.8 these two concepts, as well as the concept of the solution of the dual problem, are equivalent.

Applications of theorem 6.19 are particularly effective in cases where it is possible (in any way) to find in advance a solution $\hat{y}$ of the dual problem. Then finding solutions to the original problem is reduced to finding solutions to equation [6.79] or solutions to the system of relations [6.80] and [6.81] on the set X.

EXAMPLE 6.4.– Let us find all solutions of the problem

$$\sum_{j=1}^{n} |x_j - a_j| \to \min, \quad \sum_{j=1}^{n} x_j = 0, \qquad [6.85]$$

where $a_1, \ldots, a_n$ are fixed numbers. Let us compose the Lagrange function

$$L(x, y) = \sum_{j=1}^{n} |x_j - a_j| + y \sum_{j=1}^{n} x_j,$$

where $x = (x_1, \ldots, x_n) \in \mathbb{R}^n, \quad y \in R$. Verify that

$$\inf_{x_j \in R} (|x_j - a_j| + y x_j) = \begin{cases} y a_j, & |y| \leq 1, \\ -\infty, & |y| > 1. \end{cases}$$

Hence

$$\varphi(y) = \inf_{x \in \mathbb{R}^n} L(x, y) = \begin{cases} yA, & |y| \leq 1, \\ -\infty, & |y| > 1. \end{cases}$$

where $A = \sum_{j=1}^{n} a_j$. The dual problem has the form

$$\varphi(y) = yA \to \max, \quad -1 \leq y \leq 1.$$

We find the solution of the dual problem, which is of the form

$$\hat{y} = \operatorname{sign} A,$$

and

$$\varphi(\hat{y}) = |A|.$$

According to theorem 6.19, solutions of problem [6.85] coincide with the solutions of equation [6.79] on X, that is systems

$$\sum_{j=1}^{n} |x_j - a_j| = |A|, \quad \sum_{j=1}^{n} x_j = 0. \qquad [6.86]$$

If $A = 0$, then $x = (a_1, \ldots, a_n)$ is a unique solution of this system. Let $A \neq 0$. We will seek solutions in the form

$$x_j = a_j - \lambda_j A, \quad j = 1, \ldots, n, \qquad [6.87]$$

where $\lambda_1, \ldots, \lambda_n$ are some fixed numbers. Substituting [6.87] in [6.86], we obtain

$$\sum_{j=1}^{n} |\lambda_j| = 1, \quad \sum_{j=1}^{n} \lambda_j = 1.$$

Hence $\lambda_1 \geq 0, \ldots, \lambda_n \geq 0$. Now it is clear that all solutions of system [6.86], and, correspondingly, all solutions of problem [6.85] are described by formula [6.87], where $\lambda_1 \geq 0, \ldots, \lambda_n \geq 0$, $\sum_{j=1}^{n} \lambda_j = 1$.

It is clear from the given example that to find solutions to problem [6.62], equation [6.79] or systems [6.80] and [6.81], they should be solved exactly on X, and not on P. We point out below an important special case where the condition $\hat{x} \in X$ can be ignored, since it is performed automatically.

THEOREM 6.20.– Suppose that the assumptions of theorem 6.12 are satisfied. Let $\hat{y}$ be a solution of problem [6.67]. Let problem [6.62] have a solution. If $\hat{x}$ is the unique point of the set P satisfying one of the conditions [6.79]–[6.81], then $\hat{x}$ is the unique solution to problem [6.62].

PROOF.– Let $\bar{x}$ be a solution of problem [6.62]. Then by theorem 6.19 each of the conditions [6.79]–[6.81] is satisfied when $\hat{x}$ is replaced on $\bar{x}$. By virtue of the assumption made, this is possible only when $\bar{x} = \hat{x}$. □

Theorem 6.19 can be formulated in a more attractive form, if we use the notion of a saddle point.

DEFINITION 6.5.– *The pair $(\hat{x}, \hat{y}) \in P \times Q$ is called the saddle point of the function $L(x, y)$ on $P \times Q$ if the following relations hold true:*

$$L(\hat{x}, \hat{y}) = \min_{x \in P} L(x, \hat{y}), \qquad [6.88]$$

$$L(\hat{x}, \hat{y}) = \max_{y \in Q} L(\hat{x}, y), \qquad [6.89]$$

that is if

$$L(x, \hat{y}) \geq L(\hat{x}, \hat{y}) \geq L(\hat{x}, y)$$

for all $x \in P, y \in Q$.

The following theorem is the Kuhn–Tucker theorem as a statement about the saddle point

THEOREM 6.21.– Suppose assumptions of theorem 6.12 are fulfilled. The point $\hat{x} \in P$ is a solution of the problem [6.62] if and only if there is a vector $\hat{y} \in Q$ such that the pair $(\hat{x}, \hat{y})$ is the saddle point of the Lagrange function $L(x, y)$ on $P \times Q$.

PROOF.– Since conditions [6.80] and [6.88] coincide, we need to show that condition [6.89] is satisfied if and only if $\hat{x} \in P$ and condition [6.81] is satisfied. However, this follows from theorem 6.2 and lemma 6.3, applied to the problem of minimizing the function $-L(\hat{x}, y)$ with respect to y on Q. □

Sometimes theorems 6.19 and 6.21 are more convenient to use as a statement about conditions for simultaneous optimality of points in direct and dual problems.

THEOREM 6.22.– Let the assumptions of theorem 6.12 be fulfilled. Then:

1) points $\hat{x} \in X$ and $\hat{y} \in Y$ are solutions of the problems [6.62] and [6.67], respectively, if and only if there are satisfied the duality relation [6.79], which is equivalent to conditions [6.80] and [6.81];

2) points $\hat{x} \in P$ and $\hat{y} \in Q$ are solutions of the problems [6.80] and [6.81], respectively, if and only if the pair $(\hat{x}, \hat{y})$ is the saddle point of the Lagrange function $L(x, y)$ on $P \times Q$.

We note that in statement (1) one could take $\hat{y} \in Q$, since from the condition (6.79) it follows that $\hat{y} \in Q$. The difference between statements (1) and(2) consists in the fact that in (1) the admissibility condition $\hat{x} \in X$ is assumed immediately, and in (2) it is not assumed.

THEOREM 6.23.– (*Kuhn–Tucker theorem in sub-differential form*) Let the assumptions of theorem 6.12 be satisfied and let the functions $f(x), g_1(x), \ldots, g_k(x)$ be convex on an open convex set U, which contains P. Let us assume that the linear functions $g_{k+1}(x), \ldots, g_m(x)$ are of the form $g_i(x) = \langle a_i, x\rangle + b_i$, $i = k+1, \ldots, m$. A point $\hat{x} \in X$ is a solution of the problem [6.62] if and only if there exist vectors $\hat{y} \in Q$, $a_0 \in \partial f(\hat{x})$, $a_i \in \partial g_i(\hat{x})$, $i = 1, \ldots, k$, such that

$$\left\langle a_0 + \sum_{i=1}^{m} y_i a_i,\ x - \hat{x} \right\rangle \geq 0 \quad \text{for all} \quad x \in P, \qquad [6.90]$$

$$\hat{y}_i g_i(\hat{x}) = 0, \quad i = 1, \ldots, k.$$

PROOF.– Given theorem 6.19, it is enough to show that the relation [6.90] is equivalent to relation [6.80]. But this follows immediately from theorems 5.8 and 6.3 applied to a convex function with respect to x on U

$$L(x, \hat{y}) = f(x) + \sum_{i=1}^{m} \hat{y}_i g_i(x).$$

Note that if the functions $f(x), g_1(x), \ldots, g_k(x)$ are differentiable at point $\hat{x}$, then by theorem 5.3 relation [6.90] is transformed to [6.22], that is, statements of theorems 6.8 and 6.23 coincide in this case. □

6.6.4. *Method of perturbations*

A fruitful method for studying optimization problems is the *method of perturbations*, which is based on the approach where the initial problem is considered as one of the elements of a family of problems that depend on a parameter.

We demonstrate this method with the help of an example of the convex programming problem with inequality constraints.

$$f(x) \to \min, \quad g(x) \le 0, \quad x \in P \subset \mathbb{R}^n, \qquad [6.91]$$

where $g(x) = (g_1(x), \ldots, g_m(x))$.

We introduce perturbations in the right-hand sides of the constraints of the problem [6.91], that is, we consider a family of problems of the form

$$f(x) \to \min, \quad g(x) \le b, \quad x \in P, \qquad [6.92]$$

where $b \in \mathbb{R}^m$ is a vector-valued parameter.

We introduce the following notation:

$$X(b) = \{x \in P | g(x) \le b\}$$

is the admissible set of the problem [6.92];

$$B = \{b \in \mathbb{R}^m | X(b) \ne \emptyset\}$$

is the set of parameters b for which the problem [6.92] has admissible points;

$$F(b) = \inf_{x \in X(b)} f(x), \quad b \in B,$$

is the value of the problem [6.92];

$$\hat{Y}(b) = \{\hat{y} \in R^m_+ | F(b) \le f(x) + \langle \hat{y}, g(x) - b\rangle \quad \text{for all} \quad x \in P\}$$

is the set of Kuhn–Tucker vectors of this problem.

We will also need a sub-differential of the function $F(b)$ on B:

$$\partial F(b) = \{a \in \mathbb{R}^m | F(b') - F(b) \ge \langle a, b' - b\rangle \text{ for all } b' \in B\}.$$

THEOREM 6.24.– Let the set P be convex, let the functions $f(x)$ and $g(x)$ be convex on P, $0 \in B$, $F(0) > -\infty$ and let $\hat{Y}(0) \ne \emptyset$, that is, the value of problem [6.91] is finite and it has the Kuhn–Tucker vector (for example, satisfies one of the conditions (1)–(4) of theorem 5.3.2). Then:

1) the set B is convex;

2) the function $F(b)$ is finite, convex and monotonically non-increasing on B;

3) $\partial F(b) = -\hat{y}(b)$ for all $b \in B$.

We note that in theorem 6.15, the Kuhn–Tucker vectors of the problem [6.62] were described as solutions of the dual problem [6.67]. The assertion (3) of theorem 6.24 gives the Kuhn–Tucker vectors another characteristic: the Kuhn–Tucker vectors of the problem [6.92] taken with the opposite sign are subgradients of the values of this problem as a function of the right and left parts of restrictions.

PROOF.– 1) Let $b^1, b^2 \in B,\ \lambda \in [0,1]$. By definition of B, there exist points $x^1, x^2 \in P$ such that $g(x^1) \leq b^1, g(x^2) \leq b^2$. Take $x = \lambda x^1 + (1-\lambda)x^2$. From the convexity of the set P and the convexity of the function g, it follows that $x \in P$ and

$$g(x) \leq \lambda g(x^1) + (1-\lambda) g(x^2) \leq \lambda b^1 + (1-\lambda) b^2, \tag{6.93}$$

that is $\lambda b^1 + (1-\lambda) b^2 \in B$. Consequently, the set B is convex.

2) Consider an arbitrary vector $\hat{y} \in \hat{Y}(0)$. By definition of $\hat{Y}(0)$, we have

$$F(0) \leq f(x) + \langle \hat{y}, g(x) \rangle \quad \text{for all} \quad x \in P.$$

Since $\hat{y} \geq 0$, for arbitrary $b \in B$ and $x \in X(b)$, we have

$$F(0) \leq f(x) + \langle \hat{y}, b \rangle .$$

Then by definition of $F(b)$

$$F(0) \leq F(b) + \langle \hat{y}, b \rangle , \tag{6.94}$$

where $F(0) > -\infty$. Therefore $F(b) > -\infty$. Consequently, the function F is finite on B.

Let $b^1, b^2 \in B,\ \lambda \in [0,1]$ and let $b = \lambda b^1 + (1-\lambda) b^2$. For arbitrary $x^1 \in X(b^1)$, $x^2 \in X(b^2)$ take $x = \lambda x^1 + (1-\lambda) x^2$. Then according to [6.93], we have $x \in X(b)$. From the convexity of the function f, we get

$$F(b) \leq f(x) \leq \lambda f(x^1) + (1-\lambda) f(x^2) .$$

Therefore

$$F(b) \leq \lambda F(b^1) + (1-\lambda) F(b^2) ,$$

that is, the function F is convex on B.

Let $b^1 \in B,\ b^2 \geq b^1$. Then $X(b^1) \subset X(b^2)$. Hence, $F(b^1) \geq F(b^2)$, that is, the function F monotonically non-increasing on B.

3) Fix $b \in B$. Let $\hat{y} \in \partial F(b)$, that is

$$F(b') - F(b) \geq \langle \hat{y}, b' - b \rangle \quad \text{for all} \quad b' \in B. \qquad [6.95]$$

For arbitrary $x \in P$, take $b' = g(x)$. Then $b' \in B$ and $F(b') \leq f(x)$. Taking this into account, it follows from [6.95] that

$$F(b) \leq f(x) + \langle -\hat{y}, g(x) - b \rangle \quad \text{for all} \quad x \in P. \qquad [6.96]$$

In addition, we have $-\hat{y} \geq 0$, since the function F does not increase monotonically. So, $-\hat{y} \in \hat{y}(b)$.

Let the opposite of $-\hat{y} \in \hat{y}(b)$ hold true, that is, [6.96] holds true and let $-\hat{y} \geq 0$. Then for arbitrary $b' \in B$ and $x \in X(b')$, we get

$$F(b) \leq f(x) + \langle -\hat{y}, b' - b \rangle .$$

This implies [6.95], i.e. $\hat{y} \in \partial F(b)$. Hence, $\partial F(b) = -\hat{y}(b)$. □

We see that theorem 6.24 is simple: for its proof only knowledge of definitions is required. However, this theorem allows us to draw the problem [6.91] into a study of the rather deep facts of the convex analysis. We indicate below a number of results by specifying the assumptions of theorem 6.24.

THEOREM 6.25.– Let the set P be convex, let the functions f and g be convex on P and let $F(0) > -\infty$ and $g(\bar{x}) < 0$ for some $\bar{x} \in P$, that is, the problem [6.91] satisfies the Slater condition. Then $0 \in \operatorname{int} B$ and:

1) the function F is continuous at point $b = 0$;

2) the function F has derivative $F'(0, h)$ in any direction $h \in \mathbb{R}^m$ at point $b = 0$, besides

$$F'(0, h) = \max_{\hat{y} \in \hat{Y}(0)} \langle -\hat{y}, h \rangle ; \qquad [6.97]$$

3) the function F is differentiable at point $b = 0$ if and only if the Kuhn–Tucker vector of the problem [6.91] is unique: $\hat{Y}(0) = \{\hat{y}\}$; with $F'(0) = -\hat{y}$.

The presence of the whole spectrum of Kuhn–Tucker vectors of problem [6.91] corresponds to the case where the function $F(b)$ has a break at the point $b = 0$.

PROOF.– The inclusion of $0 \in \operatorname{int} B$ is obvious: if $g(\bar{x}) < 0$, then $g(\bar{x}) \leq b$ for all sufficiently small in the norm b. In view of this and theorem 6.24, statements (1)–(3) follow from theorems 3.20, 5.2 and 5.3 respectively. □

The assertions (2) and (3) of theorem 6.25 mean that the Kuhn–Tucker vectors, taken with the opposite sign, act as estimates of the rate of change of a problem when the right-hand sides of the restrictions change. The following theorem specifies the nature of the change (monotonically non-increasing) of the function F with the change of only one ith coordinate of the vector b.

THEOREM 6.26.– 1) If the assumptions of theorem 6.24 are satisfied and $\hat{y}_i = 0$ for some $\hat{y} \in \hat{Y}(0)$, then $F\left(\alpha e^i\right) = F(0)$ for all $\alpha > 0$ (here e^i is the ith basis vector in $\mathbb{R}^m$).

2) If the assumptions of theorem 6.25 are satisfied and $\hat{y}_i > 0$ for all $\hat{y} \in \hat{Y}(0)$, then $F\left(\alpha e^i\right) < F(0)$ for all $\alpha > 0$.

PROOF.– 1) The vector $\hat{y} \in \hat{Y}(0)$ satisfies the inequality [6.94]. If we substitute $b = \alpha e^i$ for $\hat{y}_i = 0$ in this inequality, then we obtain $F(0) \leq F\left(\alpha e^i\right)$. But for $\alpha < 0$ by the monotonic non-increasing F, we have $F\left(\alpha e^i\right) \leq F(0)$. Hence, $F(0) = F\left(\alpha e^i\right)$.

2) We take the element $\bar{y} \in \hat{Y}(0)$, on which the maximum is achieved in [6.97] with $h = e^i$:

$$F'\left(0, e^i\right) = \lim_{\alpha \to 0+} \frac{F\left(\alpha e^i\right) - F(0)}{\alpha} = -\bar{y}_i.$$

By the condition $\bar{y}_i > 0$. Then $F\left(\alpha e^i\right) < 0$. We have $F\left(\alpha e^i\right) - F(0) < 0$ for all sufficiently small $\alpha > 0$, and in general for all $\alpha > 0$, since F is monotonically non-increasing. □

We will specify another property of problem [6.91], which follows from the corresponding result of the convex analysis. Along with problem [6.92], we consider an equivalent problem of the form

$$\varphi(f(x)) \to \min, \quad g(x) \leq b, \quad x \in P, \qquad [6.98]$$

where φ is a convex differentiable increasing function on R. Let $\Phi(b)$ be the value of the problem [6.98], and let $\hat{Y}_\varphi(b)$ be the set of its Kuhn–Tucker vectors. It turns out that this set for $b = 0$ is related to the set of Kuhn–Tucker vectors of the original problem [6.91].

THEOREM 6.27.– Let the assumptions of theorem 6.25 be satisfied, and let the function φ satisfy the above conditions. Then

$$\hat{Y}_\varphi(0) = \varphi'(F(0))\,\hat{Y}(0).$$

PROOF.– It is clear that $\Phi(b) = \varphi(F(b))$. Making use of theorem 5.10 for Φ and taking into account the assertion (3) of theorem 6.24, we obtain

$$\hat{Y}_\varphi = -\partial\Phi(0) = -\varphi'(F(0))\,\partial F(0) = \varphi'(F(0))\,\hat{y}(0). \qquad \square$$

Consequently, for a monotonic transformation of the objective function, the set of Kuhn–Tucker vectors of the new problem is obtained from the Kuhn–Tucker vectors of the original problem by multiplying them by the same number.

6.6.5. *Economic interpretations of the Kuhn–Tucker vector*

The Kuhn–Tucker vector can be given various economic interpretations, depending on the type of the optimization problem and its interpretation. Two such interpretations are as follows. In the first interpretation, the Kuhn–Tucker vector appears as a vector of resource scarcity, and in the second one, it coincides with the price vector acting in the system.

Consider the maximization problem

$$f(x) \to \max, \quad g(x) \le b, \quad x \in P, \qquad [6.99]$$

where x is an n-dimensional vector of production of goods by an enterprise; P is the set of technologically possible production plans, $P \subset R^n_+$; $g(x)$ is an m-dimensional vector of resource consumption for the product x, $g(x) \ge 0$; b is a vector characterizing resources, $b > 0$ and $f(x)$ is the profit received by the enterprise from the sale of x of goods. The problem [6.99] satisfies Slater's condition: $g(0) < b$. The assumptions necessary for application of the above theory, that the set P is convex, the function f is concave and the function g is convex, also have an economic interpretation. Thus, the necessary (and sufficient when the function f is continuous) condition of concavity of f on P is the inequality

$$f(x + 2\Delta x) - f(x + \Delta x) \le f(x + \Delta x) - f(x),$$

which holds true for all $x \in P$ and $\Delta x \in \mathbb{R}^n$ such that $x + 2\Delta x \in P$. For $\Delta x \ge 0$, this inequality means that as production scales increase, the profit growth decreases (for example, due to difficulties arising in the sale of goods).

Let now $\hat{x} = \hat{x}(b)$ be a solution of the problem [6.99], and let $\Phi(b) = f(\hat{x}(b))$ be the value of the problem [6.99], that is, the optimal plan for the release of goods and the maximum profit from the sale of goods with a reserve of b resources. Let $\hat{y}$ be the Kuhn–Tucker vector of the problem [6.99], more precisely the equivalent problem

$$-f(x) \to \min, \quad g(x) - b \le 0, \quad x \in P.$$

Then, if the ith resource is not fully used, $g_i(\hat{x}) < b$, then $\hat{y}_i = 0$ by theorem 6.19. Moreover, by theorem 6.26, increasing only the ith resource cannot lead to an increase in the company's profit ($\Phi\left(b + \alpha e^i\right) = \Phi(b)$ for $\alpha > 0$). In this case, we say that the ith resource is *not deficient*. If $\hat{y}_i > 0$, then the ith resource is *deficit*: it is used completely ($g_i(\hat{x}) = b_i$). An increase in its stock leads to an increase in the profit of the enterprise ($\Phi\left(b + \alpha e^i\right) > \Phi(b)$ for $\alpha > 0$).

Further, by theorem 6.25, we have $\Phi'(b) = \hat{y}$ and $\Phi'(b, h) = \langle \hat{y}, h \rangle$.

The first formula means that if the enterprise can obtain small amounts of individual resources, the most desirable (in terms of profit growth) is the increase in the resource that corresponds to the maximum component of the vector $\hat{y}_i$, that is, such a resource is the most deficient. If it is possible to increase the number of all resources at the same time, then, according to the second formula, they should be purchased in proportions, described by the vector $h = \hat{y}$.

Thus, the Kuhn–Tucker vector of the problem [6.99] appears as an important characteristic of the insufficiency of resources that are used to produce goods. We did not take into account that the acquisition of additional resources requires certain costs on the part of the enterprise. Now let us take this circumstance into account. Let $p = (p_1, \ldots, p_m)$ be a given price vector for resources, $p > 0$. We will assume that the enterprise can both buy the necessary resources and sell the "unnecessary" part of the stock in order to maximize the overall profit, which takes into account the results of trading operations with resources. Then the activity of the enterprise is described by the problem

$$f(x) - \langle p, h \rangle \to \max, \quad g(x) \leq b + h, \quad x \in P, \quad h \geq -b, \qquad [6.100]$$

where $h = (h_1, \ldots, h_m)$ is a vector of purchase and sale of resources by the enterprise: if $h_i > 0$, then the resource i is bought, if $h_i < 0$ then the resource is sold. The condition $h \geq -b$ means that the enterprise cannot sell more resources than it has. Let $(\hat{x}, \hat{h})$ be a solution of the problem [6.100], and $\hat{y}$ be its Kuhn–Tucker vector, which corresponds to functional restriction $g(x) \leq b + h$. Then, by theorem 6.19, we have

$$L(\hat{x}, \hat{h}, \hat{y}) = \min_{x \in P, h \geq -b} L(x, h, \hat{y}),$$

where

$$L(x, h, y) = -f(x) + \langle p, h \rangle + \langle y, g(x) - b - h \rangle.$$

It is natural to assume that $\hat{h} > -b$, that is, it is not profitable for an enterprise to fully sell its resources. Then the gradient of the function $L(x, h, \hat{y})$ with respect to h at the point $(\hat{x}, \hat{h})$ vanishes:

$$L'_h(\hat{x}, \hat{h}, \hat{y}) = p - \hat{y} = 0.$$

Consequently, the Kuhn–Tucker vector of the problem [6.100] is the vector of current prices.

6.7. Exercises

1) Make sure that the set $\hat{Y}$ of the Kuhn–Tucker vectors of problem [6.62] is always convex and closed.

2) Let problem [6.62] satisfy condition (1) of theorem 6.12 (Slater's condition). Show that the set $\hat{Y}$ is bounded. Will this happen under conditions (2), (3), and (4) of theorem 6.12?

3) Formulate and prove a theorem on the existence of a subgradient of a convex function at an interior point of the domain of definition (statement 1 of theorem 5.1) from theorem 6.12; in this example, once again, we see that in assumption (4) of theorem 6.12 the assumption $\bar{x} \in \text{ri}\, P$ is essential. Hint: consider the problem $f(x) \to \min, x = \hat{x}, x \in X$.

4) Check that the convex optimization problem

$$x \to \min, \quad x^2 + \varepsilon|x| \leq 0, \quad x \in R,$$

where $\varepsilon > 0$, does not satisfy any of the conditions (1)–(4) of theorem 6.12; however, it should have the Kuhn–Tucker vector.

5) Establish the connection between the Kuhn–Tucker vector of the problem

$$f(x) \to \min, \ \ g_i(x) \leq 0, \ \ i = 1, \ldots, m, \ \ x \in P,$$

and the equivalent problem

$$f(x) \to \min, \quad \max_{i=1,\ldots,m} g_i(x) \leq 0, \ \ x \in P.$$

6) Verify that the duality relation ($\hat{f} = \hat{\varphi}$) is not satisfied for such optimization problems:

a) $x - 1 \to \min, \ \ x^2 - 1 = 0, \ \ x \in R_+$;

b) $f(x) \to \min, \ \ x^2 \leq 0, \ \ x \in R_+$, where $f(x) = \begin{cases} 1, & x = 0 \\ 0, & x > 0 \end{cases}$.

For each problem, find out the reasons for this phenomenon, referring to theorems 6.15 and 6.17.

7) General scheme of duality: let P and Q be arbitrary sets from $\mathbb{R}^n$ and $\mathbb{R}^m$, respectively, and let $L(x, y)$ be an arbitrary function on $P \times Q$. Take

$$f(x) = \sup_{y \in Q} L(x, y), \quad \varphi(y) = \inf_{x \in P} L(x, y).$$

The problem

$$f(x) \to \min, \quad x \in P,$$

is called a direct optimization problem, and the problem

$$\varphi(y) \to \max, \quad y \in Q,$$

is called a dual problem. Here infinite values are allowed for objective functions. Let

$$\hat{f} = \inf_{x \in P} f(x), \quad \hat{\varphi} = \sup_{y \in Q} \varphi(y)$$

be values of these problems, and let

$$\hat{X} = \{\hat{x} \in P | f(\hat{x}) = \hat{f}\}, \quad \hat{Y} = \{\hat{y} \in Q | \varphi(\hat{y}) = \hat{\varphi}\}$$

be sets of solutions of the indicated optimization problems. Prove the following statements:

a) $f(x) \geq \varphi(y)$ for all $x \in P$, $y \in Q$ and $\hat{f} \geq \hat{\varphi}$;

b) if $\hat{x} \in P$ and $f(\hat{x}) = \hat{\varphi}$, then $\hat{x} \in \hat{X}$;

c) if $\hat{y} \in Q$ and $\varphi(\hat{y}) = \hat{f}$, then $\hat{y} \in \hat{Y}$;

d) if $\hat{x} \in P$, $\hat{y} \in Q$ and $f(\hat{x}) = \varphi(\hat{y})$, then $\hat{x} \in \hat{X}$, $\hat{y} \in \hat{Y}$;

e) for given $\hat{x} \in P$ and $\hat{y} \in Q$ the equality $f(\hat{x}) = \varphi(\hat{y})$ is equivalent to that $(\hat{x}, \hat{y})$ is a saddle point of the function $L(x, y)$ on $P \times Q$.

8) Duality theorem: suppose, within the framework of the preceding problem, that the sets P and Q are convex, the function $L(x, y)$ is convex in x on P for each $y \in Q$ and it is concave in y on Q for each $x \in P$. Suppose that the set P (or the set Q) is closed, the set $\hat{X}$ (the set $\hat{Y}$) is non-empty and bounded, and the function $L(x, y)$ is continuous in x on P for every $y \in Q$ (in y on Q for each $x \in P$). Prove that $\hat{f} = \hat{\varphi}$.

9) Applying the scheme of reasoning from the example 6.4, solve the following optimization problems ($a_1, \ldots, a_n$, $b_1, \ldots, b_n$ are fixed numbers):

a) $\sum\limits_{j=1}^{n} |x_j - a_j| \to \min, \quad \sum\limits_{j=1}^{n} b_j x_j = 0;$

b) $\sum\limits_{j=1}^{n} \max(x_j - a_j; 0) \to \min, \quad \sum\limits_{j=1}^{n} b_j x_j \leq 0;$

c) $\sum\limits_{j=1}^{n} a_j x_j \to \min, \quad \sum\limits_{j=1}^{n} b_j |x_j| \leq 1 \quad (b_j > 0, j = 1, \ldots, n);$

d) $\sum\limits_{j=1}^{n} |x_j - a_j| \to \min, \quad \sum\limits_{j=1}^{n} x_j^2 \leq 1.$

Solutions, Answers and Hints

Solutions for Chapter 1

1) *Solution.* The function is continuous. Obviously, $S_{\max} = +\infty$. In accordance with corollary 1.1 of the Weierstrass theorem, a minimum is attained. The necessary conditions of the first order are of the form

$$\frac{\partial f}{\partial x} = 4x^3 - 4y = 0, \quad \frac{\partial f}{\partial y} = 4y^3 - 4x = 0$$

Solving these equations, we determine the critical points $(0,0)$, $(1,1)$, $(-1,-1)$. To use the second-order condition, we calculate matrices composed of the second derivatives:

$$A = \left\{ \frac{\partial^2 f}{\partial x \partial y} \right\} = \begin{pmatrix} 12x^2 & -4 \\ -4 & 12y^2 \end{pmatrix},$$

$$A_1 = A(0,0) = \begin{pmatrix} 0 & -4 \\ -4 & 0 \end{pmatrix},$$

$$A_2 = A(1,1) = A(-1,-1) = \begin{pmatrix} 12 & -4 \\ -4 & 12 \end{pmatrix}.$$

The matrix A_1 is indefinite. Therefore, the point $(0,0)$ does not satisfy the necessary conditions of the extremum of the second order. Point $(0,0) \notin \operatorname{locextr} f$. The matrix A_2 is positive definite. Therefore, in accordance with theorem 1.13, a local minimum of the problem is attained at points $(1,1)$, $(-1,-1)$.

Answer. $(0,0) \notin$ locextr; $(1,1), (-1,-1) \in$ absmin.

2) $(\ln(a + \sqrt{ab}), \ln(b + \sqrt{ab}) \in$ absmin, $S_{\max} = +\infty$.

3) $((2a - b)/3, (2b - a)/3) \in$ locmin, $a + b < 0$;

$((2a-b)/3, (2b-a)/3) \in$ locmax, $a+b>0$;

$(a,-a) \notin$ locmin, $a+b=0$.

4) $(0,0) \notin$ locextr.

5) Stationary points: (x,y), where $x = (-1)^{k+1}\pi/12 + (k+m)\pi/2, y = (-1)^{k+1}\pi/12 + (k-m)\pi/2, k, m = 0, \pm 1, \pm 2, \ldots$ Points $(x,y) \in$ locmin if k is even and m is odd. Points $(x,y) \in$ locmax if k is odd and m is even. Points $(x,y) \notin$ locextr if $k+m$ is even.

6) $(0, 2k\pi) \in$ locmin, $(-2, (2k+1)\pi) \notin$ locextr, $k = 0, \pm 1, \pm 2, \ldots$

7) $(0,0) \in$ absmin, $\{(x,y) : x^2 + y^2 = 1\} \in$ locmax.

8) $(\pm(2e)^{-1/2}, \pm(2e)^{-1/2}) \in$ locmin, $S_{\text{min}} = -1/2e$,

$(\pm(2e)^{-1/2}, \mp(2e)^{-1/2}) \in$ locmax, $S_{\text{max}} = 1/2e$;

$(0, \pm 1) \notin$ locextr, $(\pm 1, 0) \notin$ locextr.

9) $(1,1)$ is a saddle point.

10) $(2\pi/3, 2\pi/3) \in$ locmin, $S_{\text{min}} = -3\sqrt{3}/8$,

$(\pi/3, \pi/3) \in$ locmax, $S_{\text{max}} = 3\sqrt{3}/8$.

11) $(\pi/3, \pi/6) \in$ locmax, $S_{\text{max}} = 3\sqrt{3}/2$.

12) $(1,2) \in$ locmin, $S_{\text{min}} = 7 - 10\ln 2$.

13) $(-1/26, -3/26) \in$ locmin, $S_{\text{min}} = -26e^{-1/52}$,

$(1,3) \in$ locmax, $S_{\text{max}} = e^{-13}$.

14) $(1,-2)$ is a saddle point.

15) $(0,0) \in$ locmin, $S_{\text{min}} = 0$; $(-1/4, -1/2)$ is a saddle point.

16) $(0,0) \in$ locmax, $S_{\text{max}} = 1$.

17) $(a/c, b/c) \in$ locmin, $c < 0$; $(a/c, b/c) \in$ locmax, $c > 0$,

$S_{\text{min}} = -\sqrt{a^2+b^2+c^2}$, $S_{\text{max}} = \sqrt{a^2+b^2+c^2}$;

there is no extremum if $c = 0$, $a^2 + b^2 \neq 0$.

18) $(\pm a/\sqrt{3}, \mp b/\sqrt{3}) \in \text{locmin}$, $(\pm a/\sqrt{3}, \pm b/\sqrt{3}) \in \text{locmax}$,

$S_{\min} = -ab/3\sqrt{3}$, $S_{\max} = -ab/3\sqrt{3}$.

19) $(\pm 1/2, \pm 1) \in \text{locmin}$, $(0,0) \in \text{locmax}$, $S_{\min} = -9/8$, $S_{\max} = 0$;

$(0, \pm 1)$, $(\pm 1/2, 0)$ are saddle points.

20) $(1,0) \in \text{absmin}$, $S_{\min} = -1$, $S_{\max} = +\infty$.

21) $(5,2) \in \text{locmin}$, $S_{\min} = -\infty$, $S_{\max} = +\infty$.

22) $S_{\min} = -\infty$, $S_{\max} = +\infty$, $(2,3) \notin \text{locextr}$.

23) $(-4, 14) \in \text{absmin}$, $S_{\max} = +\infty$.

24) $(8, -10) \notin \text{locextr}$, $S_{\min} = -\infty$, $S_{\max} = +\infty$.

25) $(1,1) \in \text{locmax}$, $(0,0)$, $(0,3)$, $(3,0) \notin \text{locextr}$, $S_{\min} = -\infty$,

$S_{\max} = +\infty$.

26) $(-2/3, -1/3, 1) \in \text{absmin}$, $S_{\min} = -4/3$, $S_{\max} = +\infty$.

27) $(2,2,1) \in \text{absmin}$, $S_{\min} = -1$.

28) $(a/7, a/7, a/7) \in \text{locmax}$, $S_{\max} = a^7/7^7$.

29) $(24, -144, -1) \in \text{absmin}$, $S_{\min} = -6,913$.

30) $(1/2, 1, 1) \in \text{absmin}$, $S_{\min} = 4$.

31) $(-1, -2, 3) \in \text{absmin}$, $S_{\min} = -14$.

32) $(2^{1/3}, 4^{1/3}) \in \text{locmax}$, $(0,0) \notin \text{locextr}$.

33) $a \neq b$, $(a^{1/2}(a^{3/2} + b^{3/2})^{-1}, b^{1/2}(a^{3/2} + b^{3/2})^{-1}) \in \text{locmin}$;

$(a^{1/2}(a^{3/2} - b^{3/2})^{-1}, b^{1/2}(b^{3/2} - a^{3/2})^{-1}) \in \text{locmax}$;

$a = b$, $(a^{1/2}(a^{3/2} + b^{3/2})^{-1}, b^{1/2}(a^{3/2} + b^{3/2})^{-1}) \in \text{absmin}$.

34) $0 < a < 2$, $(0, -a^{1/2}) \in \text{absmin}$, $(0, a^{1/2}) \in \text{absmax}$;

$a > 2$, $(-b, -b^{-1}) \in$ locmin, $(b^{-1}, b) \in$ locmin;

$(b, b^{-1}) \in$ locmax, $(-b^{-1}, -b) \in$ locmax;

$a = 2$, $(\pm 1, \pm 1) \notin$ locextr.

35) $(-1, \pi/3 + 2k\pi) \in$ absmin, $(1, -\pi/3 + 2k\pi) \in$ absmin,

$(1, \pi/3 + 2k\pi) \in$ absmax, $(-1, -\pi/3 + 2k\pi) \in$ absmax.

36) $(-b\,\mathrm{sign}(ab)/\sqrt{a^2+b^2}, -a\,\mathrm{sign}(ab)/\sqrt{a^2+b^2}) \in$ absmin,

$(b\,\mathrm{sign}(ab)/\sqrt{a^2+b^2}, a\,\mathrm{sign}(ab)/\sqrt{a^2+b^2}) \in$ absmax,

$S_{\min} = -\sqrt{a^2+b^2}/|ab|$, $S_{\max} = \sqrt{a^2+b^2}/|ab|$.

37) $(ab^2/(a^2+b^2), a^2b/(a^2+b^2)) \in$ absmin, $S_{\min} = a^2b^2/(a^2+b^2)$.

38) $S_{\min} = \lambda_1$, $S_{\max} = \lambda_2$, $(A - \lambda)(C - \lambda) - B^2 = 0$.

39) $(\pm 2, \mp 3) \in$ absmin, $(\pm 3/2, \pm 4) \in$ absmax,

$S_{\min} = -50$, $S_{\max} = 106, 25$.

40) $(\pi/8 + k\pi/2, -\pi/8 + k\pi/2) \in$ locmin, $k = 2n+1$;

$(\pi/8 + k\pi/2, -\pi/8 + k\pi/2) \in$ locmax, $k = 2n$;

$S_{\min} = 1 + (-1)^k/\sqrt{2}$, $S_{\max} = 1 + (-1)^k/\sqrt{2}$.

41) $(8/13, 12/13) \in$ absmin, $S_{\min} = 36/13$, $S_{\max} = +\infty$.

42) $(3/25, 4/25) \in$ absmin, $S_{\min} = 1/25$, $S_{\max} = +\infty$.

43) $(1/2, 1/2) \in$ absmin, $S_{\min} = 0$, $S_{\max} = e^{1/4}$.

44) $(-1/2, 3/2) \in$ absmin, $S_{\max} = +\infty$.

45) $S_{\min} = -\infty$, $S_{\max} = +\infty$.

46) $(1/6, 1/3, 1/2) \in$ locmax; $(t, 0, 1-t) \in$ locmax, $t > 1, t < 0$;

$(t, 0, 1-t) \in$ locmin $t \in (0, 1)$; $S_{\min} = -\infty$, $S_{\max} = +\infty$.

47) $\{(1/\sqrt{6}, 1/\sqrt{6}, -2/\sqrt{6}), (1/\sqrt{6}, -2/\sqrt{6}, 1/\sqrt{6}),$

$(-2/\sqrt{6}, 1/\sqrt{6}, 1/\sqrt{6})\} \in$ absmin, $S_{\min} = -1/3\sqrt{6}$,

$\{(-1/\sqrt{6}, -1/\sqrt{6}, 2/\sqrt{6}), (-1/\sqrt{6}, 2/\sqrt{6}, -1/\sqrt{6}),$

$(2/\sqrt{6}, -1/\sqrt{6}, -1/\sqrt{6})\} \in$ absmax, $S_{\min} = 1/3\sqrt{6}$.

48) $\{(0,0,\pm 1), (0,\pm 1, 0), (\pm 1, 0, 0)\} \in$ absmin,$S_{\min} = 0$;

$(\pm 1/\sqrt{2}, 0, \pm 1/\sqrt{2}) \in$ absmax, $S_{\max} = (a-c)^2/4$;

$(0, \pm 1/\sqrt{2}, \pm 1/\sqrt{2}) \notin$ locextr,$(\pm 1/\sqrt{2}, \pm 1/\sqrt{2}, 0) \notin$ locextr.

49) $(1/2, 1/2, 1/2) \in$ locmin; $\{(x, y, z) : (x-1/2)^2 + (y-1/2)^2 = 2,$

$x + y + z = -1/2\} \in$ absmin.

50) $(-1/3, 2/3, -2/3) \in$ absmin; $S_{\min} = -3$;

$(1/3, -2/3, 2/3) \in$ absmax; $S_{\max} = 3$.

51) $S_{\max} = a^{m+n+p} m^m n^n p^p / (m+n+p)^{m+n+p}$,

$x/m = y/n = z/p = a/(m+n+p)$.

52) $(0, 0, \pm c) \in$ absmin; $S_{\min} = c^2$; $(\pm a, 0, 0) \in$ absmax; $S_{\max} = a^2$.

53) $(a/6, a/6, a/6) \in$ absmax; $S_{\max} = (a/6)^6$.

54) $(1, 1, 1) \in$ absmax; $S_{\max} = 2$.

55) $(\pi/6, \pi/6, \pi/6) \in$ absmax; $S_{\max} = 1/8$.

56) $(0, 1) \in$ absmin, $S_{\min} = 1/e - 1$; $(1, 0) \in$ absmax, $S_{\max} = e - 1$.

61) $\{(-1/\sqrt{3}, -1/\sqrt{3}, -1/\sqrt{3}), (1/\sqrt{3}, 1/\sqrt{3}, -1/\sqrt{3}),$

$(1/\sqrt{3}, -1/\sqrt{3}, 1/\sqrt{3}), (-1/\sqrt{3}, 1/\sqrt{3}, 1/\sqrt{3})\} \in$ absmin,

$\{(1/\sqrt{3}, 1/\sqrt{3}, 1/\sqrt{3}), (1/\sqrt{3}, -1/\sqrt{3}, -1/\sqrt{3}),$

$(-1/\sqrt{3}, -1/\sqrt{3}, 1/\sqrt{3})\} \in$ absmax,

$S_{\min} = -1/3\sqrt{3}$, $S_{\max} = 1/3\sqrt{3}$.

62) $(-2, 0, 7) \in$ absmin, $S_{\max} = +\infty$.

63) $(0,0,0) \in \text{absmin}$, $S_{\min} = 0$; $(0,12,0) \in \text{absmax}$, $S_{\max} = 576$.

64) $(0,1,0) \in \text{absmin}$, $S_{\min} = 0$, $S_{\max} = +\infty$.

65) $(2/7, 174/35, -24/5) \in \text{locmin}$, $S_{\min} = -\infty$; $(1,0,3) \in \text{locmax}$,

$S_{\max} = +\infty$.

67) $(0,\ldots,0) \in \text{absmin}$, $S_{\min} = 0$,

$(\pm n^{-1/4},\ldots,\pm n^{-1/4}) \in \text{absmax}$, $S_{\max} = \sqrt{n}$.

68) $(0,\ldots,0) \in \text{absmin}$, $S_{\min} = 0$, $(\pm 1, 0,\ldots,0),\ldots,$

$(0,\ldots,0,\pm 1) \in \text{absmax}$, $S_{\max} = 1$.

77) *Solution.* 1. Formalization:

$$f(x) = x(8-x)(8-2x) \to \sup, \quad 0 \le x \le 4.$$

According to the Weierstrass theorem, the problem has a solution $\hat{x}$. Obviously, $\hat{x} \neq 0$ and $\hat{x} \neq 4$, since $f(0) = f(4) = 0$.

2. By Fermat's theorem, $f'(\hat{x}) = 0$.

3. $f'(\hat{x}) = 0 \Longleftrightarrow 3x^2 - 24x + 32 = 0 \Rightarrow \hat{x} = 4 \pm 4/\sqrt{3}$.

Answer. One number is $4 - 4/\sqrt{3}$, the other is $4 + 4/\sqrt{3}$.

78) An isosceles triangle.

79) The point E is the midpoint of the segment BC.

80) The center of weight of a fixed face.

81) $t^2 - 1/3 \in \text{absmin}$.

82) $t^3 - 3t/5 \in \text{absmin}$.

83) $\hat{p} = (\hat{p}_1,\ldots,\hat{p}_n) \in \text{absmax}$, $\hat{p}_1 = \ldots = \hat{p}_n = 1/n$.

84) Square.

85) The right triangle.

86) The height of the cylinder is equal to $2/\sqrt{3}$.

87) The height of the cone is equal to $4/3$.

88) The height of the cone is equal to $4/3$.

89) Cube.

90) The right tetrahedron.

91) The right triangle.

92) The right polygon.

93) The right polygon.

94) *Solution.* 1. Formalization:

$$\sqrt{1 - a^2 \sin^2(\varphi)} \sin(\varphi) \to \sup, \quad 0 \le \varphi \le \pi/2,$$

the angle CEB is denoted φ, and the area of the quadrilateral inscribed in a circle is equal to half the product of diagonals by the sine of the angle between them. After replacing $a \sin(\varphi) = \sqrt{z}$, we obtain the following problem:

$$f(z) = z(1 - z) \to \sup, \quad 0 \le z \le a^2.$$

According to the Weierstrass theorem, the problem has a solution $\hat{z}$.

2. If $0 < \hat{z} < a^2$, then by Fermat's theorem $f'(\hat{z}) = 0$.

3. $f'(\hat{z}) = 0 \Longleftrightarrow \hat{z} = 1/2$.

4. Since the problem has a solution, then one of the stationary points 0, $1/2$, a^2 gives the absolute maximum of the problem. Comparing values of the function $f(z)$ at these points, we find a solution.

Answer. If $0 \le a \le 1/\sqrt{2}$, then $\hat{z} = a^2$, that is $\hat{\varphi} = \pi/2$. If $1/\sqrt{2} \le a \le 1$, then $\hat{z} = 1/2$, that is $\hat{\varphi} = \arcsin(a\sqrt{2})-1$.

95) The center of the inscribed circle.

96) *Solution.* 1. Formalization:

$$f(x_1, x_2) = |x_1 - x_2|^2 + |x_1 - e|^2 + |x_2 - e|^2 \to \sup;$$

$$|x_1|^2 = |x_2|^2 = 1, \quad e = (1, 0), \quad x_i \in R^2, \quad i = 1, 2,$$

$$|x| = \sqrt{x_1^2 + x_2^2}.$$

The Lagrange function

$$\mathcal{L} = \lambda_0 f(x_1, x_2) + \lambda_1 |x_1|^2 + \lambda_2 |x_2|^2.$$

2. The necessary conditions of the extremum

$$\mathcal{L}_{x_1} = 0 \Longleftrightarrow \lambda_0(x_1 - x_2 + x_1 - e) + \lambda_1 x_1 = 0,$$
$$\mathcal{L}_{x_2} = 0 \Longleftrightarrow \lambda_0(x_2 - x_1 + x_2 - e) + \lambda_2 x_2 = 0.$$

3. If $\lambda_0 = 0$, then $\lambda_1 = \lambda_2 = 0$. All Lagrange multipliers are zeros. Let $\lambda_0 = -1$. Then

$$(2 - \lambda_1)x_1 = x_2 + e, \quad (2 - \lambda_2)x_2 = x_1 + e \Rightarrow$$
$$((2 - \lambda_1)(2 - \lambda_2) - 1)x_1 = e(3 - \lambda_2),$$

So, $x_1 = \pm e$, or $\lambda_1 = \lambda_2 = 3 \Rightarrow x_1 + x_2 = -e$. Therefore, (1) $x_1 = x_2 = \pm e$, (2) $x_1 = -x_2 = e$ and (3) $x_1 = (-1/2, \sqrt{3}/2)$, $x_2 = (-1/2, -\sqrt{3}/2)$.

4. Since the set of admissible elements is compact, according to the Weierstrass theorem, the solution of the problem exists. The maximum of the functional gives the third extreme.

Answer. The right triangle.

97) *Solution*. 1. Let M be a given point within the angle AOB, points C, D lie on the rays OA and OB, respectively, and $M \in [CD]$. Draw a straight line parallel to OA through M. We denote by F the point of intersection of this line with the line OB. Suppose that OF is a and FD is x. The area of the triangle OCD is equal to $k(a+x)^2/x$, where k is a coefficient of proportionality. The problem

$$S(x) = k(a+x)^2/x \to \inf, \quad x \geq 0$$

has a solution because $S(x)$ is continuous and $S(x) \to \infty$ as $x \to +\infty$.

2. Write the necessary condition of the extremum $S'(\hat{x}) = 0$.

3. Solve the equation $S'(\hat{x}) = 0 \Longleftrightarrow \hat{x} = a$.

4. Since the stationary point is one, then $\hat{x} \in \text{absmin}$.

Answer. A straight line drawn so that its segment between the sides of the angle is divided by a given point into two equal parts.

98) Through a point, draw a circle (larger radius) that touches the sides of the corner and a segment tangent to the circle.

99) A quadrilateral inscribed in a circle.

100) *Solution*. 1. Formalization

$$V(R,h) = \pi h^2(R - h/3) \to \sup, \quad 2\pi Rh = a, \quad 0 \le h \le 2R,$$

where R is the radius of the ball, h is the height of the segment and a is the lateral surface area. Excluding R, we get

$$V(h) = \frac{ah}{2} - \frac{\pi k^3}{3} \to \sup, \quad 0 \le h \le \sqrt{a/\pi}.$$

2. Write down the necessary condition of the extremum $V'(\hat{h}) = 0$.

3. Solve the equation $V'(\hat{h}) = 0 \Longleftrightarrow \hat{h} = \sqrt{a/2\pi}$.

4. According to the Weierstrass theorem, there exists a solution of the problem. Because $\hat{h} \neq 0$ and $\hat{x} \neq \sqrt{a/\pi}$ since $V(0) = 0$, $V(\sqrt{a/\pi}) = a\sqrt{a}/6\sqrt{\pi} < V(\sqrt{a/2\pi}) = a\sqrt{a}/3\sqrt{2\pi}$, then $a = 2\pi\hat{h}^2$, whence $\hat{h} = R$.

Answer. The hemisphere has the largest volume.

101) If points A, B lie on different sides of the line l, then C is the intersection point of lines AB and l. Let points A and B lie on one side of the line l. Find a point A' symmetric to the point A with respect to the line l. The intersections of the lines $A'B$ and l determine the point C.

102, 103) The top of the tetrahedron is projected into the center of the circle inscribed in the base.

104) The right tetrahedron.

105) $x_0 = (x_1 + x_2 + x_3)/3$.

106) $x_0 = \left(\sum\limits_{i=1}^{N} m_i x_i\right) \Big/ \sum\limits_{i=1}^{N} m_i.$

107) Let

$$\hat{x} = \left(\sum_{i=1}^{N} m_i x_i\right) \Big/ \sum_{i=1}^{N} m_i.$$

If $|\hat{x}| \le 1$, then $x_0 = \hat{x}$; if $|\hat{x}| > 1$, then $x_0 = \hat{x}/|\hat{x}|$.

108) Let

$$\hat{x} = \left(\sum_{i=1}^{N} m_i x_i\right) \Big/ \sum_{i=1}^{N} m_i.$$

If $|\hat{x}| = 0$, then x_0 is arbitrary; if $|\hat{x}| \neq 0$, then $x_0 = \hat{x}/|\hat{x}|$.

109) From the point (ξ_1, ξ_2) to the ellipse $x_1^2/a_1^2 + x_2^2/a_2^2 = 1$, $a_1 > a_2$ four normals can be drawn if this point lies inside the astroid

$$(\xi_1 a_1)^{2/3} + (\xi_2 a_2)^{2/3} = (a_1^2 - a_2^2)^{2/3};$$

three normals if it lies on the astroid (except for the vertices). From other points, two normals can be drawn.

110) From the point (ξ_1, ξ_2) to the parabola $y = ax^2$, $a > 0$, three normals can be drawn if the point is located above the curve

$$\xi_2 = 3 \cdot 2^{-4/3} a^{-1/3} \xi_1^{2/3} + 2^{-1} a^{-1};$$

two normals if it lies on this curve (except for the point $(0, 2^{-1}a^{-1})$). From other points, one normal can be drawn.

111) From the point (ξ_1, ξ_2), we can draw three normals to the near branch of the hyperbola and one to the far branch if

$$(\xi_1 a_1)^{2/3} - (\xi_2 a_2)^{2/3} > (a_1^2 + a_2^2)^{2/3};$$

two normals to the near and one to the far if

$$(\xi_1 a_1)^{2/3} - (\xi_2 a_2)^{2/3} = (a_1^2 + a_2^2)^{2/3}$$

(except for the points $(0, (a_1^2 + a_2^2)/a_1)$). From other points, one normal to each branch can be drawn.

112) The distance from the point $\hat{x} = (\hat{x}_1, \ldots, \hat{x}_n)$ to the hyperplane $\sum_{i=1}^{n} a_i x_i = b$ equals $\left|\sum_{i=1}^{n} a_i x_i - b\right| / \left(\sum_{i=1}^{n} a_i^2\right)^{1/2}$.

113) *Solution.* 1. Suppose that the hyperplane is defined by the equation $\langle a, x\rangle - b = 0$, where $\langle a, x\rangle$ is the scalar product of vectors a, x. The formalized problem has the form

$$\|x - x_0\|^2 \to \inf, \quad \langle a, x\rangle - b = 0, \quad a \neq 0.$$

We compose the Lagrange function

$$\mathcal{L} = \lambda_0 \|x - x_0\|^2 + \lambda \langle a, x \rangle.$$

2. Write down the necessary condition $2\lambda_0(\hat{x} - x_0) + \lambda a = 0$.

3. If $\lambda_0 = 0$, then $a = 0$. There are no solutions. Let $\lambda_0 = 1/2$. Then

$$\hat{x} = x_0 - \lambda a, \quad \lambda = (\langle a, x_0 \rangle - b)/\langle a, a \rangle.$$

4. We make sure that $\hat{x} \in \text{absmin}$,

$$S_{\min} = (\langle a, x_0 \rangle - b)/\langle a, a \rangle.$$

Answer. The distance equals $|\langle a, x_0 \rangle - b)|/\|a\|$.

114) The distance from the point $\hat{x}$ to the straight line $at + b$, $a, b \in R^n$ equals

$$(\|\hat{x} - b\|^2 - (\langle \hat{x} - b, a \rangle / \|a\|^2)^{1/2}.$$

115) $\hat{x} = -a/\|a\| \in \text{absmin}$, $l(\hat{x}) = \|a\|$.

116) Rectangle sides: $\sqrt{2}a$, $\sqrt{2}b$.

117) The sides of the parallelepiped: $2a/\sqrt{3}, 2b/\sqrt{3}, 2c/\sqrt{3}$.

118) *Solution.* We investigate the constrained optimization problem

$$\sum_{i=1}^{n} |x_i|^p \to \sup; \quad \sum_{i=1}^{n} |x_i|^q = a^q, \quad 1 < p < q, a > 0.$$

1. The set of admissible elements is compact. The objective function is continuous. By the Weierstrass theorem, a solution to the problem exists.

2. We compose the Lagrange function

$$\mathcal{L} = \lambda_0 \sum_{i=1}^{n} |x_i|^p + \lambda \left(\sum_{i=1}^{n} |x_i|^q - a^q \right).$$

3. We write the first-order necessary condition

$$\hat{\mathcal{L}}_{x_i} = 0 \Longleftrightarrow \lambda_0 p |\hat{x}_i|^{p-1} \operatorname{sign}(\hat{x}_i) + \lambda q |\hat{x}_i|^{q-1} \operatorname{sign}(\hat{x}_i) = 0, \ i = 1, \dots, n.$$

4. If $\lambda_0 = 0$, then $\hat{x} = 0$ will not be an admissible element of the problem. Let $\lambda_0 = -1$. Then $\hat{x}_i = 0$ or $|\hat{x}_i| = (\lambda q/p)^{1/(p-q)}$.

5. The functional reaches its maximum at a critical point. Let a critical point have k non-zero coordinates. These coordinates are as follows: $|\hat{x}_k| = ak^{-1/q}$. So,

$$S_{\max}(a) = a^p \max_{1 \le k \le n} k^{1-p/q} = a^p n^{1-p/q}.$$

Using the solution of the problem, we prove the inequality. Let $p > 1$ and let $\sum_{i=1}^{n} |x_i|^q = a^q$. Then

$$\left(\sum_{i=1}^{n} |x_i|^p/n\right)^{1/p} = n^{-1/p}\left(\sum_{i=1}^{n} |x_i|^p\right)^{1/p} \le n^{-1/p}(S_{\max}(a))^{1/p}$$

$$= n^{-1/p} a n^{1/p-1/q} = a n^{-1/q} = \left(\sum_{i=1}^{n} |x_i|^q/n\right)^{1/q}.$$

If $p = 1$, then we can verify the inequality by passing to the limit in inequalities with $p > 1$.

Let $0 < p < 1$ and $y_i = |x_i|^p$. Then

$$\left(\sum_{i=1}^{n} |x_i|^p/n\right)^{1/p} = \left(\sum_{i=1}^{n} |y_i|/n\right)^{1/p}$$

$$\le \left(\left(\sum_{i=1}^{n} |y_i|^{q/p}/n\right)^{p/q}\right)^{1/q} = \left(\sum_{i=1}^{n} |x_i|^q/n\right)^{1/q}.$$

If $p < q < 0$, then $-q < -p$ and we can use the proved inequalities

$$\left(\sum_{i=1}^{n} |x_i|^p/n\right)^{1/p} = \left(\sum_{i=1}^{n} |x_i^{-1}|^{-p}/n\right)^{-(-1/p)}$$

$$\le \left(\sum_{i=1}^{n} |x_i^{-1}|^{-q}/n\right)^{-(-1/q)} = \left(\sum_{i=1}^{n} |x_i|^q/n\right)^{1/q}.$$

Let $p < 0 < q$. Then

$$\lim_{p \to 0}\left(\sum_{i=1}^{n} |x_i|^p/n\right)^{1/p} = \left(\prod_{i=1}^{n} |x_i|\right)^{1/n},$$

$$\left(\sum_{i=1}^{n} |x_i|^p/n\right)^{1/p} \le \left(\prod_{i=1}^{n} |x_i|\right)^{1/n} \le \left(\sum_{i=1}^{n} |x_i|^q/n\right)^{1/q}.$$

119) The inequality is deduced in the same way as in problem 118.

120) The solution is the same as in problem 118.

121) *Solution.* We investigate the constrained optimization problem

$$\sum_{i=1}^{n} x_i a_i \to \sup; \quad \sum_{i=1}^{n} |x_i|^p = b^p, \quad p > 1, b > 0.$$

1. The set of admissible elements is compact. The objective function is continuous. By the Weierstrass theorem, a solution to the problem exists.

2. We compose the Lagrange function

$$\mathcal{L} = \lambda_0 \sum_{i=1}^{n} x_i a_i + \lambda \sum_{i=1}^{n} |x_i|^p.$$

3. We write the necessary condition for the extremum

$$\hat{\mathcal{L}}_{x_i} = 0 \Longleftrightarrow \lambda_0 a_i + \lambda p |\hat{x}_i|^{p-1} \operatorname{sign}(\hat{x}_i) = 0, \; i = 1, \ldots, n.$$

4. If $\lambda_0 = 0$ and $\lambda \neq 0$, then $\hat{x} = 0$ will not be an admissible element of the problem. Let $\lambda_0 = -1$. Then

$$\hat{x}_i = \mu |a_i|^{p'-1} \operatorname{sign}(a_i), \quad 1/p + 1/p' = 1.$$

Since $\sum\limits_{i=1}^{n} |x_i|^p = b^p$, then $\mu = \left(\sum\limits_{i=1}^{n} |a_i|p' \right)^{-1/p}$.

5. We found only one critical point. Therefore $\hat{x} \in \operatorname{absmax}$.

$$S_{\max}(b) = b \left(\sum_{i=1}^{n} |a_i|^{p'} \right)^{-1/p'}.$$

So,

$$\sum_{i=1}^{n} a_i x_i \leq S_{\max} \left(\left(\sum_{i=1}^{n} |x_i|^p \right)^{1/p} \right) = \left(\sum_{i=1}^{n} |x_i|^p \right)^{1/p} \left(\sum_{i=1}^{n} |x_i|^{p'} \right)^{1/p'}.$$

122) Hint: investigate the constrained optimization problem

$$\sum_{i=1}^{n} |x_i + y_i|^p \to \sup; \; \sum_{i=1}^{n} |x_i|^p = a^p, \; \sum_{i=1}^{n} |y_i|^p = b^p, \; p > 1, \; a > 0, \; b > 0.$$

Solutions for Chapter 2

2) Let $x^1 \in X$ and $x^2 \in X$ and let

$$x(\lambda) = \lambda x^1 + (1-\lambda)x^2.$$

Let there exist $\lambda_0 \in (0,1)$ such that $x(\lambda_0) \notin X$. Let

$$\lambda_1 = \sup\{\lambda | x(\lambda) \in X, 0 \leq \lambda \leq \lambda_0\},$$

$$\lambda_2 = \inf\{\lambda | x(\lambda) \in X, \lambda_0 \leq \lambda \leq 1\}.$$

Since X is a closed set, we have $x(\lambda_1) \in X$ and $x(\lambda_2) \in X$ and $\lambda_1 \leq \lambda_0 \leq \lambda_2$. We get a contradiction since for points $x(\lambda_1) \in X$ and $x(\lambda_2) \in X$ the condition of the problem is not satisfied. As an example of unclosed set, we can take the set of all rational numbers.

3) a) Let $x \in X_1$ and $y \in X_1$, that is $x_1^2 \leq x_2$ and $y_1^2 \leq y_2$, and let $\lambda \in (0,1)$. Then

$$(\lambda x_1 + (1-\lambda)y_1)^2 = \lambda^2 x_1^2 + (1-\lambda)^2 y_1^2 + 2\lambda(1-\lambda)x_1 y_1$$

$$\leq \lambda^2 x_2 + (1-\lambda)^2 y_2 + 2\lambda(1-\lambda)\sqrt{x_2 y_2}$$

$$\leq \lambda^2 x_2 + (1-\lambda)^2 y_2 + \lambda(1-\lambda)(x_2 + y_2)$$

$$= \lambda x_2 + (1-\lambda)y_2.$$

This means that $\lambda x + (1-\lambda)y \in X_1$. So X_1 is convex.

3) c) Let $x \in X_3$ and $y \in X_3$. It follows from exercise 2 that it is sufficient to prove that $0.5x + 0.5y \in X_3$. This is true if

$$\frac{\sin x_1 + \sin y_1}{2} \leq \sin \frac{x_1 + y_1}{2}.$$

It is equivalent to

$$\left(\sin \frac{x_1}{2} - \sin \frac{y_1}{2}\right)\left(\cos \frac{x_1}{2} - \cos \frac{y_1}{2}\right) \leq 0.$$

The last inequality holds true since $x_1 \in [0,\pi]$ and $y_1 \in [0,\pi]$.

5) Let $x \in X$ and $y \in X$. Then

$$\left\langle A\left(\frac{1}{2}x + \frac{1}{2}y\right), \frac{1}{2}x + \frac{1}{2}y\right\rangle = \frac{1}{4}\langle Ax, x\rangle + \frac{1}{4}\langle Ay, y\rangle + \frac{1}{2}\langle Ax, y\rangle$$

$$= \frac{1}{2}\langle Ax, x\rangle + \frac{1}{2}\langle Ay, y\rangle - \frac{1}{4}\langle A(x-y), x-y\rangle$$

$$\leq \frac{1}{2}\langle Ax, x\rangle + \frac{1}{2}\langle Ay, y\rangle \leq \alpha.$$

It follows from exercise 2 that the set X is convex. Let $\alpha = 0$. Then X is also a cone and $\langle Ax, x\rangle = 0$ for all $x \in X$. Therefore, for $x \in X, y \in X$ and $\lambda \in \mathbb{R}$ we have $x + y \in X$ and $\lambda x \in X$, that is, X is a linear subspace.

6) X is a cone. To prove that X is a convex cone, we use the Cauchy–Bunyakovsky inequality

$$(x_1 + y_1)^2 - 2(x_1 + y_1)(x_3 + y_3) + (x_2 + y_2)^2$$

$$(x_1^2 - 2x_1x_3 + x_2^2) + (y_1^2 - 2y_1y_3 + y_2^2) + 2(x_1y_1 + x_2y_2 - y_1x_3 - x_1y_3)$$

$$\leq 2\left(\sqrt{x_1^2 + x_2^2}\sqrt{y_1^2 + y_2^2} - y_1x_3 - x_1y_3\right)$$

$$\leq 2\left(2\sqrt{x_1x_3y_1y_3} - y_1x_3 - x_1y_3\right) \leq 0.$$

That is, $x + y \in X$. So X is a convex cone.

10) Let X_1 be an open set and let X_2 be an arbitrary set. We can write that the sum of sets

$$X_1 + X_2 = \cup_{x \in X_2}(X_1 + x)$$

and use the fact that the union of an arbitrary number of open sets is open.

11) Let

$$X = \{x \in \mathbb{R}^2 | x_1x_2 \geq 1, x_1 > 0\}$$

and

$$Y = \{x \in \mathbb{R}^2 | x_1x_2 \geq 1, x_1 < 0\}.$$

The sum $X + Y$ is not closed.

12) Let X be a cone determined in exercise 2.6 and let $Y = \{x \in \mathbb{R}^2 | x_1 = 0, x_2 = 0\}$. The sum $X + Y$ is not closed.

13) Projection $P(X)$ of the set $X \subset \mathbb{R}^n \times \mathbb{R}^m$ on the space of the first n coordinates is an image of the mapping $P(x, y) = x, x \in \mathbb{R}^n, y \in \mathbb{R}^m$. So it is convex. The cone determined in exercise 2.6 and its projection on XOY is an example of a closed convex set such that its projection is not closed.

16) Let X_1, X_2 be convex sets in $\mathbb{R}^n$ and let

$$Y = \bigcup_{x \in X_2} \bigcup_{\lambda \geq 1} (\lambda X_1 + (1 - \lambda)x).$$

Let $x^1 \in Y, x^2 \in Y, \alpha \in (0,1)$ and $x = \alpha x^1 + (1-\alpha)x^2$. Then $x^i = \lambda y^i + \mu_i z^i$, $y^i \in X_1$, $z^i \in X_2$, $\lambda_i \geq 1$, $\mu_i = 1 - \lambda_i$, $i = 1, 2$. Consider $\beta = \alpha\lambda_1 + (1-\alpha)\lambda_2 \geq 1$ and $\gamma = 1 - \beta$. In the case where $\beta > 1$, we have

$$x = \beta\left(\frac{\alpha\lambda_1}{\beta}y^1 + \frac{(1-\alpha)\lambda_2}{\beta}y^2\right) + \gamma\left(\frac{\alpha\mu_1}{\gamma}z^1 + \frac{(1-\alpha)\mu_2}{\gamma}z^2\right).$$

In this case, $x \in Y$. In the case where $\beta = 1$, we have $\lambda_1 = 1$, $\lambda_2 = 1$. So $x = \alpha y^1 + (1-\alpha)y^2 \in X_1 \subset Y$.

17) Let $x \in X, y \in X, \alpha \in (0,1)$. Then

$$x = \sum_{i=1}^{m} \lambda_i x^i, \quad y = \sum_{i=1}^{m} \mu_i y^i,$$

for some $\lambda, \mu \in \Lambda$, $x^i \in X_i$, $y^i \in X_i$, $i = 1, 2 \ldots, m$. Let us introduce the notation $\eta = \alpha\lambda + (1-\alpha)\mu \in \Lambda$. For any $i = 1, 2 \ldots, m$ such that $\eta_i > 0$ denote

$$z^i = \frac{\alpha\lambda_i}{\eta_i}x^i + \frac{(1-\alpha)\mu_i}{\eta_i}y^i \in X_i.$$

In the case $\eta_i = 0$, we have $\lambda_i = 0$, $\mu_i = 0$. In this case, we can take $z^i \in X_i$ arbitrary. Finally, we have

$$\alpha x + (1-\alpha)y = \sum_{i=1}^{m} \eta_i z^i \in X.$$

As an example that shows that the condition $\Lambda \subset \mathbb{R}^m_+$ is significant here consider the case where $m = 1$, $X_1 = \mathbb{R}^n_+$, $\Lambda = \mathbb{R}$. In this case, $X = \mathbb{R}^n_+ \cup (-\mathbb{R}^n_+)$ is not convex.

18) We prove the theorem by induction on dimension n of the space $\mathbb{R}^n$. For $n = 1$, the statement is obvious. Suppose that it is true in the case $\mathbb{R}^{n-1}, n \geq 2$. Consider $x \in X, y \in Y$. We have to prove that $x + y \in X$. Denote $u = \min\{x, y\}$, $v = \max\{x, y\}$. We have $x + y = u + v, u \leq v$. So it is sufficient to prove that $u + v \in X$. In the case where $u = 0$, we have $u + v = v \in X$. Let $u \neq 0$. Take

$$\lambda = \min_{j:u_j>0} \frac{v_j}{u_j} = \frac{v_{j_0}}{u_{j_0}} \geq 1 > 0.$$

Let $j_0 = n$. Then $\lambda u \leq v$ and $\lambda u_n = v_n$. For this reason

$$(1+\lambda)u_j \leq u_j + v_j \leq \left(1 + \frac{1}{\lambda}\right)v_j, \quad j = 1, 2, \ldots, n-1,$$

$$(1+\lambda)u_n = v_n = \left(1 + \frac{1}{\lambda}\right)v_n.$$

Denote by Y projection of X on the space of first $n-1$ coordinates. Y is a cone in the space $\mathbb{R}_+^{n-1}$. It is closed with respect to operations $\min\{x,y\}$ and $\max\{x,y\}$. It is a convex cone because of the assumption of the method of induction. Vectors $(u_1,\ldots,u_{n-1})$ and $(v_1,\ldots,v_{n-1})$ belong to Y. So their sum also belongs to Y. In other words, there exists a vector $w \in X$ such that $w_j = u_j + v_j, j = 1,2,\ldots,n-1$. Moreover

$$u + v = \max\{w, \left(1+\frac{1}{\lambda}\right) v\} \in X$$

in the case where $w_n \le u_n + v_n$, and

$$u + v = \min\{w, \left(1+\frac{1}{\lambda}\right) v\} \in X$$

in the case where $w_n > u_n + v_n$,

The statement is proved.

As an example that shows that conditions of the theorem are significant here consider the following sets:

$$X_1 = \{x \in \mathbb{R}^2 | 0 \le x_1 \le 1, 0 \le x_2 \le 1, \quad \text{or } x_1 = x_2 > 1\}$$

$$X_2 = \mathbb{R}_+^n \cup (-\mathbb{R}_+^n).$$

20) Take a point $x \in X$ and construct sequences $\{a^k\}$ and $\{y^k\}$ as follows. Take $a^1 \in A$. If a point a^k is constructed, then we take $a^{k+1} \in A$ and $y^k \in X_2$ such that

$$x + a^k = y^k + a^{k+1}, \quad k = 1,2,\ldots$$

Then we take sum of the first s of these equalities and divide it by s. We get

$$x + \frac{a^1}{s} = x^s + \frac{a^{k+1}}{s}, \quad x^s = \frac{1}{s}\sum_{k=1}^{s} y^s.$$

We have $x^s \in X_2$ since X_2 is convex. Passing to the limit and taking into account the boundedness of A, we get $x^s \to x$ as $s \to \infty$, that is $x \in \overline{X_2}$. Therefore $X_1 \subset \overline{X_2}$.

21) a) $\text{conv}X_1 = \left\{x \in \mathbb{R}^2 \left| x_1^2 \le x_2 \right.\right\}$; $\text{cone}X_1 = \left\{x \in \mathbb{R}^2 \left| x_2 > 0 \right.\right\} \cup \{0\}$.

b) $\text{conv}X_2 = \left\{x \in \mathbb{R}^2 \left| x_1^2 \le x_2, x_1 > 0 \right.\right\} \cup \{0\}$; $\quad \text{cone}X_2 = \text{int}\,\mathbb{R}_+^2 \cup \{0\}$.

c) $\text{conv}X_3 = \text{cone}X_2 = \mathbb{R}^2$.

d) $\text{conv}\,X_4 = \left\{x \in \mathbb{R}^2 \left| \sin(x_1) \ge x_2 \ge 0,\ 0 \le x_1 \le \pi \right.\right\}$;

$\text{cone}X_4 = \{x \in \mathbb{R}^2 \,|x_1 > x_2 \geq 0\} \cup \{0\}.$

e) $\text{conv}\, X_5 = \{x \in \mathbb{R}^2 \,|e^{x_1} \leq x_2\};$

$\text{cone}X_5 = \{x \in \mathbb{R}^2 \,|e^{x_1} \leq x_2, x_2 > 0\} \cup \{0\}.$

22) Let X be an open set. We have $X = \text{int}\, X \subset \text{int}\,(\text{conv}\, X)$ and the set $\text{int}\,(\text{conv}\, X)$ is convex. Therefore, $\text{conv}\, X \subset \text{int}\,(\text{conv}\, X)$. Since the set $\text{int}\,(\text{conv}\, X) \subset \text{conv}\, X$ we have $\text{int}\,(\text{conv}\, X) = \text{conv}\, X$. So the set $\text{conv}\, X$ is open.

23) No, it is not true. Examples are as follows:

a) $X_1 = \{x \in \mathbb{R}^2 \,|x_1^2 = x_2\};$

b) $X_2 = \{x \in \mathbb{R}^2 \,|x_1^2 = x_2,\ x_1 \geq 0\};$

c) $X_3 = \{x \in \mathbb{R}^2 \,|x_1^2 + (x_2 - 1)^2 \leq 1\}.$

24) The formula is not correct for arbitrary X. An example is $X = \{(1,0),(0,1)\}$.

25) The sum $X_1 + X_2$ is convex combination of the points

$(3,-1),(5,1),(1,3),(4,4),(1,0),(3,2),(-1,4),(2,5).$

Convex combination of these points form a hexagon with vertexes at the indicated points, except points $(1,3),(3,2)$.

26) The sum $X_1 + X_2$ is a hexagon with vertexes at points

$(-5,3),(-2,4),(2,4),(3,-1),(1,-4),(-4,1).$

27) We will prove the first relation. Inclusion $\text{conv}\, A\,(X) \subset A\,(\text{conv}\, X)$ is obvious since $A\,(\text{conv}\, X)$ is a convex set. Let $y \in A\,(\text{conv}\, X)$, that is $y = Ax$, where

$$x = \sum_{i=1}^{m} \lambda_i x^i,\ x^i \in X,\ \lambda_i \geq 0,\ i = 1, 2 \ldots, m,\ \sum_{i=1}^{m} \lambda_i = 1.$$

Since $y = Ax = \sum_{i=1}^{m} \lambda_i A x^i$, we have $y \in \text{conv}\, A(X)$.

28) The last equality is not true for arbitrary sets. As an example consider the sets $X_1 = \{(1,0)\}$ and $X_2 = \{(0,1)\}$.

29) The first formula is obvious. To prove the second one, we have

$$\text{aff}\left(\bigcup_{i=1}^{m} X_i\right) = \text{aff}\left(\text{cone}\left(\bigcup_{i=1}^{m} X_i\right)\right)$$

$$= \mathrm{aff}\left(\sum_{i=1}^{m} \mathrm{cone}\, X_i\right) = \sum_{i=1}^{m} \mathrm{aff}\, (\mathrm{cone}\, X_i) = \sum_{i=1}^{m} \mathrm{aff}\, X_i.$$

The last relation is not true for arbitrary sets. As an example consider the sets $X_1 = \{(1,0)\}$ and $X_2 = \{(0,1)\}$.

30) Note that for all $i = 1, 2, \ldots, m$

$$Y_i \subset X_i = \mathrm{conv}\, Y_i \subset \mathrm{conv}\left(\bigcup_{i=1}^{m} Y_i\right).$$

31) Let

$$A = \mathrm{conv}\left(\bigcup_{i=1}^{m} X_i\right), \quad B = \mathrm{conv}\left(\bigcup_{i=1}^{m} Y_i\right) + \mathrm{cone}\left(\bigcup_{i=1}^{m} Z_i\right).$$

Since $X_i \subset B$, then $A \subset B$ and $\overline{A} \subset \overline{B}$. Let $x \in B$, that is

$$x = \sum_{i=1}^{m}\sum_{j=1}^{k_i} \mu_{ij} y^{ij} + \sum_{i=1}^{m}\sum_{s=1}^{l_i} \eta_{is} z^{is}, \quad \sum_{i=1}^{m}\sum_{j=1}^{k_i} \mu_{ij} = 1,$$

for some $y^{ij} \in Y_i, z^{is} \in Z_i, \mu_{ij} \geq 0, \eta_{is} \geq 0, i = 1, 2, \ldots, m;\ j = 1, 2, \ldots, k_i;\ s = 1, 2, \ldots, l_i$. Take

$$\lambda_i = \sum_{j=1}^{k_i} \mu_{ij}, \quad i = 1, 2, \ldots, m.$$

If $\lambda_i > 0$ for all $i = 1, 2, \ldots, m$, we can write

$$x = \sum_{i=1}^{m} \lambda_i \left(\sum_{j=1}^{k_i} \mu_{ij} y^{ij} + \sum_{s=1}^{l_i} \eta_{is} z^{is}\right) = \sum_{i=1}^{m} \lambda_i x_i,$$

where $x_i \in X_i$. Therefore $x \in A$. If $\lambda_i = 0$ for some $i = 1, 2, \ldots, m$, we can change μ_{ij} a little in the sum $\sum_{i=1}^{m}\sum_{j=1}^{k_i} \mu_{ij} = 1$ to get $\lambda_i > 0$ for all $i = 1, 2, \ldots, m$ and get $x \in \overline{A}$. So $B \subset \overline{A}$ and $\overline{B} \subset \overline{A}$.

32) $(1,2), (2,3), (2,4), (2,5)$.

33) $(1,3,5), (1,4,5), (2,3,5), (2,4,5), (3,5,6), (4,5,6)$.

33) $(1,2,3), (1,3,4), (1,4,5), (1,3,5), (2,4,5)$.

34) It is necessary to solve the linear programming problem

$$\sum_{j=1}^{6} \lambda_j \to \min, \ \sum_{j=1}^{6} \lambda_j x_j = x^0, \ \lambda_j \geq 0, j = 1, 2, \ldots, 6.$$

Applying the simplex method, we get $\lambda_1 = 1$, $\lambda_2 = 0$, $\lambda_3 = 0$, $\lambda_4 = 2$, $\lambda_5 = 1$, $\lambda_6 = 0$.

37) a) Let points $x^0, x^1, \ldots, x^m$ be affinely independent and let their convex hull be $X = \operatorname{conv}\{x^0, x^1, \ldots, x^m\}$. Then $\operatorname{aff} X = \operatorname{aff}\{x^0, x^1, \ldots, x^m\}$. Therefore, $L = \operatorname{Lin}(\operatorname{aff} X) = \operatorname{aff} X - x^0$ is a linear subspace generated by vectors $x^1 - x^0, \ldots, x^m - x^0$. These vectors are linearly independent. Therefore, $\dim X = \dim L = m$.

b) Let $P = \operatorname{conv}\{x^0, x^1, \ldots, x^m\}$ be a simplex of maximal dimension in X, that is $x^0, x^1, \ldots, x^m$ is a system of affinely independent points in X. Then $X \subset \operatorname{aff} P$ and $\operatorname{aff} X \subset \operatorname{aff} P$. Therefore $\operatorname{aff} X = \operatorname{aff} P$ and $\dim X = \dim P$.

39) Suppose that

$$\sum_{i=1}^{m-1} \alpha_i x^i + \alpha x = 0.$$

We can write

$$0 = \sum_{i=1}^{m-1} \alpha_i x^i + \alpha \sum_{i=1}^{m} \lambda_i x^i = \sum_{i=1}^{m-1} (\alpha_i + \alpha\lambda_i)\, x^i + \alpha\lambda_m x^m.$$

Since $x^1, \ldots, x^m$ are linearly independent, we have $\alpha\lambda_m = 0$ and $\alpha_i + \alpha\lambda_i = 0$, $i = 1, 2, \ldots, m-1$. Since $\lambda_m \neq 0$, we have $\alpha = 0$ and $\alpha_i = 0$ for all $i = 1, 2, \ldots, m-1$. This means that $x^1, \ldots, x^{m-1}, x$ are linearly independent.

41) Let x^1 and x^2 be two projections of a point a on X, that is

$$\|x^1 - a\| = \|x^2 - a\| = \inf_{x \in X} \|x - a\|.$$

Then the triangle with vertices at points a, x^1 and x^2 is isosceles. Any internal point $x \in (x^1, x^2)$ has shorter distance to the vertex a than points x^1 and x^2. At the same time, $x \in X$ since X is convex.

42) From the inequality indicated in lemma 2.10, we have

$$\|\hat{x} - a\|^2 \leq \langle \hat{x} - a, x - a \rangle \leq \|\hat{x} - a\| \|x - a\| \quad \text{for all} \quad x \in X.$$

Therefore

$$\|\hat{x} - a\| \leq \|x - a\| \quad \text{for all} \quad x \in X.$$

This means that $\hat{x}$ is a projection of a point a on X.

43) Denote $b = \pi_X(a^1) - \pi_X(a^2)$. From lemma 2.10, we have

$$\langle a^1 - \pi_X(a^1), -b\rangle \leq 0, \quad \langle a^2 - \pi_X(a^2), b\rangle \leq 0.$$

From these inequalities, we get

$$\|b\|^2 \leq \langle a^1 - a^2, b\rangle \leq \|a^1 - a^2\| \|b\|.$$

44) The matrix that determines the quadratic function f in the left side of the inequality is non-negative definite. Therefore, the set X is convex. The function f is differentiable. That is why the supporting hyperplane is a tangential hyperplane to the boundary $\partial X = \{x | f(x) = 25\}$ at point x^0. The equation of this tangential hyperplane is $\langle f'(x^0), x - x^0\rangle = 0$. Inserting given numbers, we get the tangential hyperplane $4x_1 + 9x_2 + 4x_3 = 25$.

45) The hyperplane is of the form

$$H_{p,\beta} = \{x \in \mathbb{R}^3 | \langle p, x\rangle = \beta, p = x^0 - x, \beta = \langle p, \hat{x}\rangle\},$$

where $\hat{x}$ is a projection of x^0 onto X. Since $x^0 \notin X$, then $\hat{x}$ lies on the boundary of X. It is a solution of the optimization problem

$$\left(x_1 + \frac{5}{4}\right)^2 + \left(x_2 - \frac{5}{16}\right)^2 + \left(x_3 - \frac{15}{16}\right)^2 \to \min, \quad x_3 = x_1^2 + x_2^2.$$

So $\hat{x} = (-1, \frac{1}{4}, \frac{17}{16})$ and the equation of the hyperplane is $-32x_1 + 8x_2 - 16x_3 = 17$.

46) The parameter k has the indicated property if and only if the point $x^0 = (2, 1, 3)$ is a solution of the linear programming problem

$$\langle p, x\rangle \to \max, \quad x \in X.$$

The analysis of the problem gives us the following inequality $\frac{36}{23} \leq k \leq \frac{84}{37}$.

47) Making use of the simplex method and analysis of solution of the corresponding linear programming problem, we get $k \geq 1$.

48) $H_{p,\beta} = \{(x_1, x_2) | 3x_1 + 2x_2 = 6\sqrt{2}\}$.

49) Let a hyperplane $H_{p\beta}$ support a set X at point $a \in \overline{X}$, that is

$$\langle p, x\rangle \leq \langle p, a\rangle = \beta, \; \forall x \in X$$

and

$$a = \sum_{i=1}^{m} \lambda_i x^i, \ \lambda_i > 0, i = 1, \ldots, m, \ \sum_{i=1}^{m} \lambda_i = 1.$$

If there exists i such that $\langle p, x^i \rangle < \beta$, then

$$\beta = \langle p, a \rangle = \sum_{i=1}^{m} \lambda_i \langle p, x^i \rangle < \beta \sum_{i=1}^{m} \lambda_i = \beta = \langle p, a \rangle.$$

We get a contradiction.

53) $\beta = \sup_{x \in X} \langle p, x \rangle$.

56) Let X be a strictly convex set. Suppose that there exist a hyperplane $H_{p,\beta}$ and two different points a^1 and a^2 such that $X \in H^-_{p,\beta}$, $a^1 \in H_{p,\beta}$, $a^2 \in H_{p,\beta}$. Then on the one hand, $a = 0.5a^1 + 0.5a^2 \in H_{p,\beta}$ and, on the other hand, $a = 0.5a^1 + 0.5a^2 \in \operatorname{int} X \in H^-_{p,\beta}$. Contradiction. Next, let X be not strictly convex, that is there exist two different points a^1 and a^2 and $\lambda \in (0, 1)$ such that $a = \lambda a^1 + (1 - \lambda)a^2 \notin \operatorname{int} X$, so $a \in \partial X$. Then there exists a hyperplane $H_{p,\beta}$, which is supporting X at point a and $a^1 \in H_{p,\beta}$, $a^2 \in H_{p,\beta}$. That is why $X \cap H_{p,\beta}$ contains more than one point.

57) Let $x \in \operatorname{conv} X \cap \partial(\operatorname{conv} X)$. By the Carathéodory theorem, the point x can be represented in the form

$$x = \sum_{i=0}^{n} \lambda_i x^i, \ x^i \in X, \lambda_i \geq 0, i = 1, 2, \ldots, m, \ \sum_{i=0}^{n} \lambda_i = 1.$$

We will suppose that $\lambda_i > 0, i = 1, 2, \ldots, m$ since in other cases the theorem is proved. Let $H_{p,\beta}$ be a supporting hyperplane to the set X at point x. Then $x^i \in H_{p,\beta}$ for all $i = 1, 2, \ldots, m$. Therefore n vectors $x^1 - x^0, \ldots, x^n - x^0$ lie in the linear subspace $H_{p,0}$ and they are linearly dependent and vectors $x^0, x^1, \ldots, x^n$ are affinely dependent. So the point x can be represented as a convex combination of not more than n points from X.

58) No. Consider the set

$$X = \{x \in \mathbb{R}^2 | x_1^2 + (x_2 + 1)^2 \leq 1\}.$$

59) Denote $Y = \mathbb{R}^n \backslash X$. The sets X and Y are separated. There exists a hyperplane $H_{p,\beta}$ such that $X \subset H^-_{p,\beta}$ and $Y \subset H^+_{p,\beta}$. The set Y is open. So $Y = \operatorname{int} Y \subset \operatorname{int} H^+_{p,\beta}$. Therefore $H^-_{p,\beta} = \mathbb{R}^n \backslash \operatorname{int} H^+_{p,\beta} \subset \mathbb{R}^n \backslash Y = X$, that is $X = H^-_{p,\beta}$.

60) Let X_1 and X_2 be convex sets in $\mathbb{R}^n$, with $\mathrm{ri}\, X_1 \cap \mathrm{ri}\, X_2 = \emptyset$. These sets are proper separated, that is there exist a vector p and a number β such that

$$\langle p, x^1 \rangle \leq \beta \leq \langle p, x^2 \rangle, \quad \text{for all } x^1 \in X_1,\ x^2 \in X_2,$$

$$\langle p, \hat{x}^1 \rangle < \langle p, \hat{x}^2 \rangle, \quad \text{for some } \hat{x}^1 \in X_1,\ \hat{x}^2 \in X_2,$$

In the case where the set X_1 is bounded, we can take the hyperplane $H_{p,\alpha}$ with $\alpha = \sup_{x \in X_1} \langle p, x^1 \rangle$.

In the case where the set X_1 is a cone, we can take the hyperplane $H_{p,0}$ since in this case $0 \in X_1$ so $0 \leq \beta$. On the other hand, $\langle p, \lambda x \rangle \leq \beta$ for all $\lambda > 0$ so $\langle p, x \rangle \leq 0$.

61) Let

$$X_i - \{x \subset \mathbb{R}^n | A_i x \leq b^i,\},\quad i = 1, 2.$$

Then condition $X_1 \cap X_2 = \emptyset$ means that the system

$$A_1 x \leq b^1, \quad A_2 x \leq b^2$$

is incompatible. In this case, there exist vectors $q^1 \geq 0$ and $q^2 \geq 0$ such that

$$q^1 A_1 + q^2 A_2 = 0, \quad \langle q^1, b^1 \rangle + \langle q^2, b^2 \rangle < 0.$$

Take

$$p = q^1 A_1 = -q^2 A_2,\ \beta_1 = \langle q^1, b^1 \rangle,\ \beta_2 = -\langle q^2, b^2 \rangle.$$

Then for all $x \in X_1$, we have

$$\langle p, x \rangle = \langle q^1 A_1, x \rangle = \langle q^1, A_1 x \rangle \leq \langle q^1, b^1 \rangle = \beta_1.$$

In the same way, we show that $\langle p, x \rangle \geq \beta_2$ for all $x \in X_2$. Finally, we have

$$\sup_{x \in X_1} \langle p, x \rangle \leq \beta_1 < \beta_2 \leq \inf_{x \in X_2} \langle p, x \rangle.$$

This means that X_1 and X_2 are strongly separated.

63) Consider in the space $\mathbb{R}^{nm}$ the set $X = X_1 \times X_2 \cdots \times X_m$ and the set $Y = \{y = (z, \ldots, z), z \in \mathbb{R}^n\}$. It follows from the conditions of the problem that these sets are convex and do not intersect. Then there exists a non-zero vector $p = (p^1, \ldots, p^m) \in \mathbb{R}^{nm}$ such that for all $x = (x^1, \ldots, x_m) \in X$ and $y = (z, \ldots, z) \in Y$, the following relations hold true:

$$\langle p, x \rangle = \sum_{i=1}^{m} \langle p^i, x^i \rangle \leq \langle p, y \rangle = \sum_{i=1}^{m} \langle p^i, z \rangle = \left\langle \sum_{i=1}^{m} p^i, z \right\rangle.$$

Since z can take any value from $\mathbb{R}^n$, then $\sum_{i=1}^{m} p^i = 0$ and $\sum_{i=1}^{m} \langle p^i, x^i \rangle \leq 0$ for all $x^i \in X_i$, $i = 1, 2, \ldots, m$. Therefore $\beta_i = \sup_{x_i \in X_i} \langle p^i, x^i \rangle$ is finite and $\sum_{i=1}^{m} \beta_i \leq 0$. So the sets $X_1, \ldots, X_m$ are separated.

Solutions for Chapter 3

1) Prove this property by contradiction. Suppose that there exist points $x^1,\ x^2 \in X$ and a point $\hat{x} = \lambda x^1 + (1-\lambda)x^2 \in X$, $\lambda \notin [0,1]$ such that

$$f(\hat{x}) < \lambda f(x^1) + (1-\lambda) f(x^2).$$

Consider the case where $\lambda < 0$. Then

$$x^2 = \beta x^1 + (1-\beta)\hat{x}, \quad \beta = -\frac{\lambda}{1-\lambda} \in (0,1).$$

From the convexity of the function $f(x)$ on X, we have

$$\begin{aligned}
&f(x^2) \le \beta f(x^1) + (1-\beta) f(\hat{x}) \\
&< \beta f(x^1) + (1-\beta)(\lambda f(x^1) + (1-\lambda) f(x^2)) \\
&= (\beta + (1-\beta)\lambda) f(x^1) + (1-\beta)(1-\lambda) f(x^2)) = f(x^2),
\end{aligned}$$

which is impossible. Reasoning similarly we will come to contradiction in the case where $\lambda > 1$.

4) Let a continuous function f be midpoint convex, that is, for all $x_1 \in X$, $x_2 \in X$

$$f\left(\frac{x_1 + x_2}{2}\right) \le \frac{f(x_1) + f(x_2)}{2}.$$

We first prove that from the midpoint convexity it follows that for points $x_1, x_2, \ldots, x_n \in X$ we have the inequality

$$f\left(\frac{x_1 + x_2 + \cdots + x_n}{n}\right) \le \frac{f(x_1) + f(x_2) + \cdots + f(x_n)}{n}.$$

To prove the indicated inequality, we first prove it for integer numbers of the form $n = 2^k$. We will use induction on k. For $k = 1$, inequality follows from the definition. Let it be proved for $n = 2^k$ and prove it for $n = 2^{k+1}$. Write the following inequalities:

$$\begin{aligned}
&f\left(\frac{x_1 + x_2 + \cdots + x_{2^k} + x_{2^k+1} + \cdots + x_{2^{k+1}}}{2^{k+1}}\right) \\
&= f\left(\frac{1}{2}\left(\frac{x_1 + x_2 + \cdots + x_{2^k}}{2^k}\right) + \frac{1}{2}\left(\frac{x_{2^k+1} + \cdots + x_{2^{k+1}}}{2^k}\right)\right) \\
&\le \frac{1}{2}\left(f\left(\frac{x_1 + x_2 + \cdots + x_{2^k}}{2^k}\right) + f\left(\frac{x_{2^k+1} + \cdots + x_{2^{k+1}}}{2^k}\right)\right) \\
&\le \frac{1}{2}\left(\frac{f(x_1) + f(x_2) + \cdots + f(x_{2^k})}{2^k} + \frac{f(x_{2^k+1}) + \cdots + f(x_{2^{k+1}})}{2^k}\right)
\end{aligned}$$

$$= \frac{f(x_1) + f(x_2) + \cdots + f(x_{2^k}) + f(x_{2^k+1}) + \cdots + f(x_{2^{k+1}})}{2^{k+1}}$$

Therefore, the inequality holds true for the numbers of the form $n = 2^k$. Consider now an arbitrary n and find the least number of the form 2^k such that $2^k \geq n$. Let $x_1, x_2, \ldots, x_n \in X$. Consider the sequence

$$y_1 = x_1, y_2 = x_2, \ldots, y_n = x_n, \quad y_{n+1} = \cdots y_{2^k} = x_1 + x_2 + \cdots + x_n/n.$$

For this sequence, we have the inequality

$$f\left(\frac{y_1 + y_2 + \cdots + y_{2^k}}{2^k}\right) \leq \frac{f(y_1) + f(y_2) + \cdots + f(y_{2^k})}{2^k}.$$

Making use of the introduced y_i, we have

$$f\left(\frac{x_1 + x_2 + \cdots + x_n + (2^k - n)(x_1 + \cdots + x_n)/n}{2^k}\right)$$

$$\leq \frac{f(x_1) + f(x_2) + \cdots + f(x_n) + (2^k - n) f\left(\frac{x_1 + \cdots + x_n}{n}\right)}{2^k}.$$

This means that

$$f\left(\frac{x_1 + x_2 + \cdots + x_n}{n}\right)$$

$$\leq \frac{f(x_1) + f(x_2) + \cdots + f(x_n)}{2^k} + \frac{2^k - n}{2^k} f\left(\frac{x_1 + \cdots + x_n}{n}\right)$$

or

$$\left(1 - \frac{2^k - n}{2^k}\right) f\left(\frac{x_1 + x_2 + \cdots + x_n}{n}\right)$$

$$\leq \frac{f(x_1) + f(x_2) + \cdots + f(x_n)}{2^k}.$$

Dividing by $n2^{-k}$, we get

$$f\left(\frac{x_1 + x_2 + \cdots + x_n}{n}\right) \leq \frac{f(x_1) + f(x_2) + \cdots + f(x_n)}{n}.$$

For rational numbers $r_1 = \frac{p}{q}$ and $r_2 = 1 - \frac{p}{q} = \frac{q-p}{q}$, we have

$$f(r_1 x_1 + r_2 x_2) = f\left(\frac{p x_1 + (q-p) x_2}{q}\right)$$

$$= f\left(\frac{\sum_{i=1}^{p}(x_1) + \sum_{j=1}^{q-p}(x_2)}{q}\right)$$

$$\leq \frac{\sum_{i=1}^{p} f(x_1)}{q} + \frac{\sum_{j=1}^{q-p} f(x_2)}{q}$$

$$= \frac{p}{q} f(x_1) + \frac{q-p}{q} f(x_2) = r_1 f(x_1) + r_2 f(x_2).$$

So the inequality

$$f(\lambda x_1 + (1-\lambda)x_2) \leq \lambda f(x_1) + (1-\lambda) f(x_2)$$

holds true whenever $\lambda \in \mathbb{Q}, 0 \leq \lambda \leq 1$.

By virtue of the continuity of the function f, we conclude that the inequality holds true whenever $\lambda \in \mathbb{R}, 0 \leq \lambda \leq 1$.

7) a) $f(x) = \frac{1}{x}$, b) $f(x) = \exp(x)$.

8) b) Let $x_\lambda = \lambda x^1 + (1-\lambda)x^2$, $\lambda \in [0,1]$. To prove that

$$f(x_\lambda) \geq \lambda f(x^1) + (1-\lambda) f(x^2)$$

we have to prove

$$\varphi(x^1)\varphi(x^2) \leq \varphi(x_\lambda)(\lambda \varphi(x^1) + (1-\lambda)\varphi(x^2)).$$

The last inequality is equivalent to the inequality

$$\lambda(1-\lambda)[(\varphi(x^1))^2 + (\varphi(x^2))^2]^2 \geq 0$$

that holds true for all $\lambda \in [0,1]$.

13) a) The matrix

$$f''(x_1, x_2) = \begin{vmatrix} 4 & -4 \\ -4 & 0 \end{vmatrix}$$

is indefinite. Therefore $f(x_1, x_2)$ is neither convex nor concave.

b) Definiteness of the matrix

$$f''(x_1, x_2) = \exp(-x_1 - 3x_2) \begin{vmatrix} x_1 - 2 & 3(x_1 - 1) \\ 3(x_1 - 1) & 9x_1 \end{vmatrix}$$

depends on x_1. Therefore $f(x_1, x_2)$ is neither convex nor concave.

c) The matrix

$$f''(x_1, x_2) = \begin{vmatrix} -4 & 4 \\ 4 & -6 \end{vmatrix}$$

has eigenvalues $\lambda_1 = -5 + \sqrt{17} < 0$ and $\lambda_1 = -5 - \sqrt{17} < 0$. So it is negative definite. Therefore $f(x_1, x_2)$ is concave.

d) The matrix

$$f''(x_1, x_2, x_3) = \begin{vmatrix} 4 & 2 & -5 \\ 2 & 2 & 0 \\ -5 & 0 & 4 \end{vmatrix}$$

is indefinite. Therefore $f(x_1, x_2, x-3)$ is neither convex nor concave.

14) $f(x) = x^2(x^2 - 1)$, $f'(x) = 4x^3 - 2x$, $f''(x) = 12x^2 - 2 \geq 0$ if $x^2 \geq 1/6$. Thus $f(x)$ is convex over $S_1 = \{x : x \geq 1/\sqrt{6}\}$ and over $S_2 = \{x : x \leq -1/\sqrt{6}\}$. Moreover, since $f'(x) > 0$ for $x > 1/\sqrt{6}$ and for $x < -1/\sqrt{6}$ and thus $f(x)$ lies strictly above the tangent line for all $x \in S_1$ and for all $x \in S_2$, the function $f(x)$ is strictly convex over S_1 and S_2. For all remaining values of x, the function $f(x)$ is strictly concave.

17) Hint: for each s the function $f(sx)$ is convex in x, so $\int_0^1 f(sx)ds$ is convex.

19) a) Hint: apply the results of the preceding exercise to the function $F(x) = f(x) + \frac{1}{2}\|x\|^2$.

b) Hint: find a function f on $\mathbb{R}^2$ for which this system of equations can be represented in the form $f'(x) = -\lambda x$.

24) Hint: the gradient of f is $f'(X) = -X^{-2}$.

References

Alekseev, V.M., Galeev, E.M., Tikhomirov, V.M. (1984). *Collection of Problems in Optimization*. Nauka, Moscow (in Russian).

Alekseev, V.M., Tikhomirov, V.M., Fomin, S.V. (1987). *Optimal Control*. Consultants Bureau, New York.

Ashmanov, S.A. and Timokhov, A.V. (1991). *Optimization Theory in Problems and Exercises*. Fizmatlit, Moscow (in Russian).

Aubin, J.-P. (1998). *Optima and Equilibria: An Introduction to Nonlinear Analysis*. Springer, Berlin.

Bertsekas, D., Nedic, A., Ozdaglar A.E. (2003). *Convex Analysis and Optimization*. Athena Scientific, Belmont.

Borwein, J.M. and Lewis, A.S. (2000). *Convex Analysis and Nonlinear Optimization*. Springer-Verlag, New York.

Boyd, S. and Vandenberghe, L. (2004). *Convex Optimization*. Cambridge University Press, Cambridge.

Clarke, F.H. (1983). *Optimization and Nonsmooth Analysis*. Wiley-Interscience, New York.

Hiriart-Urruty, J.-B. (1998). *Optimisation et analyse convexe*. Presses Universitaires de France, Paris.

Hiriart-Urruty J.-B. and Lemaréchal C. (1993). *Convex Analysis and Minimization Algorithms*. Springer-Verlag, Heidelberg.

Hiriart-Urruty J.-B. and Lemaréchal C. (2001). *Fundamentals of Convex Analysis*. Springer-Verlag, Berlin.

Ioffe, A.D. and Tihomirov, V.M. (1979). *Theory of Extremal Problems*. North-Holland Pub. Co., Amsterdam and New York.

Nesterov, Y. (2004). *Introductory Lectures on Convex Optimization. A Basic Course.* Kluwer Academic Publishers, Boston.

Pshenichnyj, B.N. (1980). *Convex Analysis and Extremal Problems.* Nauka, Moscow (in Russian).

Rockafellar, R.T. (1970). *Convex Analysis.* Princeton University Press, Princeton.

Sukharev, A.G., Timokhov, A.V., Fedorov, V.V. (2005). *A Course in Optimization Methods.* Fizmatlit, Moscow (in Russian).

Van Tiel, J. (1984). *Convex Analysis. An Introductory Text.* John Wiley and Sons, Chichester.

Index

D

E

F, H

I, J

K

L

M, N

O, P

Q, R

S

T, V, W, Y, Z

Printed in the USA/Agawam, MA
September 16, 2026

910661.024